CONTENTS

Quantum Mysticism	1
A Personal Note from the Author	3
Chapter 1: The Quantum Veil – Unfolding the Cosmic Tapestry	6
Chapter 2: Light and Shadow – The Dance of Duality	11
Chapter 3: Entangled Souls – The Mystery of Quantum Entanglement	18
Chapter 4: The Observer's Gaze – Perception as Reality	25
Chapter 5: The Void and the Plenum – Embracing Emptiness	32
Chapter 6: The Dance of Probability – Embracing Uncertainty	39
Chapter 7: Superpositions of Self – Embracing Paradox	47
Chapter 8: Time Unbound – Beyond Linear Perception	54
Chapter 9: Cosmic Symphony – The Music of the Spheres	61
Chapter 10: The Alchemist's Dream –	68

Transmutation in the Quantum Field

Chapter 11: The Unseen Dimensions – Beyond the Physical Realm ... 76

Chapter 12: The Divine Uncertainty – Heisenberg and the Limits of Knowing ... 83

Chapter 13: Waves of Consciousness – Mind as Quantum Energy ... 90

Chapter 14: The Secret Code – Sacred Geometry and Quantum Patterns ... 97

Chapter 15: The Cosmic Mirror – Reflections of Self in the Quantum World ... 104

Chapter 16: The Threshold of the Infinite – Quantum Infinity and Spiritual Awakening ... 111

Chapter 17: Quantum Alchemy – The Transformative Power of Intention ... 118

Chapter 18: The Silence Between – Embracing the Space of the in-Between ... 125

Chapter 19: The Soul's Spin – Mysticism and Quantum Spin ... 140

Chapter 20: The Mystic's Eye – Perceiving Beyond the Veil ... 148

Chapter 21: The Art of Collapse – Creation Through Collapse of the Wave Function ... 155

Chapter 22: Hidden Variables – The Search for Deeper Meaning ... 171

Chapter 23: Beyond Boundaries – The Enigma of Non-Locality ... 178

Chapter 24: The Dreaming Cosmos – Reality as a Cosmic Dream ... 184

Chapter 25: The Quantum Pilgrimage – The Seeker's Odyssey in a Boundless Cosmos	191
Chapter 26: The Field of All Possibilities – Embodying the Quantum Mind	197
Chapter 27: Bridging Science and Mysticism – The Spiritual Scientist	203
Chapter 28: Enlightenment in the Quantum Age – Wisdom of a Unified Reality	209
Chapter 29: The Final Observer – Revisiting the Self and the Infinite	217
Chapter 30: Epilogue – The Eternal Mystery	224
References:	229
Acknowledgements	233
Copyright Information	235
Disclaimer	236

QUANTUM MYSTICISM

The Spiritual Implications Of Modern Physics

Dr Bhaskar Bora

DR BHASKAR BORA

A PERSONAL NOTE FROM THE AUTHOR

My journey, once marked by certainty and driven by purpose, has transformed in ways I could never have anticipated. It is no longer about grand achievements or the pursuit of external success, but about the quiet, tender moments that reveal the true essence of life—moments of love, care, and presence. What you hold in your hands is not just a collection of words, but a testament to resilience, a story woven from the delicate threads of struggle, acceptance, and ultimately, renewal.

There was a time when my life flowed with the grace of a symphony, every note in perfect harmony. As a doctor, my days were filled with the pulse of life itself—offering hope, easing suffering, and healing with steady hands. The white coat I wore wasn't just a symbol of my profession; it embodied my very identity; an outward reflection of the healer I believed I was destined to be. The lives I touched, the people I helped—it all gave profound meaning to my existence.

But life, in its mysterious and unpredictable ways, had other plans. In one swift, unforeseen moment, the world I knew unravelled. First came the spinal cord injury, stripping away the physical strength I had relied upon. Then, the shadow of cancer darkened the horizon, a stark reminder of life's fragility. The world of medicine, where I once found so much joy and purpose, suddenly slipped away, leaving a vast emptiness in its wake—a silence where once there had been meaning.

Gone were the bustling corridors of the hospital, replaced by the quiet solitude of my home. No longer a "Doctor," I found myself standing at the edge of an uncertain future, my hands—once so steady with the knowledge of healing—trembling with questions I wasn't ready to face. Without the title, without the work that had defined me for so long, who was I? What was left of me when everything I had known was no longer within reach?

In that silence, in the stillness of a life interrupted, I began to uncover something unexpected. The role of a disabled husband and father, once a distant concept, became my new reality—one that held unexpected grace.

What began as an effort to nurture my relationships, to find solace in this new world, slowly evolved into a profound inward journey.

I found healing in the spiritual—a rhythm of meditation, reading, and reflection that allowed me to rediscover the parts of myself I thought were lost. As I immersed myself in books, audiobooks, and hours of research, I began to understand that this new chapter of my life was not an ending, but a rebirth. The solitude of these years, the quiet hours of writing and reflection, gave birth to the very pages you hold in your hands now.

It is with deep gratitude that I share these words with you, knowing that they carry with them not just knowledge, but a piece of my soul. I hope that these reflections and insights offer you a fresh perspective on life and perhaps some nourishment for your own journey.

We cannot control what the universe throws at us, but how we react to those curveballs defines who we are and what we make of our lives.

CHAPTER 1: THE QUANTUM VEIL – UNFOLDING THE COSMIC TAPESTRY

In the infinite chiaroscuro of existence, where shadows of certainty dissolve into the nebulous glow of possibility, the quantum veil hangs like an ethereal curtain—an enigma spun from the threads of the cosmos itself. Beyond this shimmering scrim lies a universe not bound by the linear march of time or the rigid laws of causality but alive with the pulsations of an ineffable symphony. Each quark, each photon is a note in this vast opus, vibrating with secrets that hum beneath the threshold of our perception.

The veil does not merely conceal; it seduces, coaxing us to unravel its mysteries even as it drapes itself in new layers of paradox. To stand at its threshold is to be enveloped by the sublime terror of infinitude, a recognition that the familiar is but a shadow cast by the luminous unknown. Beneath its folds, reality is not static but a seething, metamorphic expanse—a tapestry of endless becoming, where matter and meaning dance an eternal pas de deux.

The Liminal Cadence of Reality

The quantum realm defies our desire for absolutes. Here, the axioms of classical physics crumble into a shimmering flux, a liminal cadence where particles are waves and waves are particles—a reality where being is contingent upon the act of seeing. This paradoxical domain blurs the boundaries between existence and nonexistence, revealing a universe that is less a machine and more a mirage, coalescing only when consciousness casts its gaze.

In this space of perpetual flux, time is unthreaded, its linearity an illusion that collapses under scrutiny. The quantum veil ripples with potentials, each thread of its weave representing a possibility yearning to be realized. Like whispered promises from a dream half-remembered, these probabilities shimmer with the allure of the almost-real, awaiting the alchemy of observation to crystallize into form.

Consciousness: The Sculptor of Reality

At the heart of this enigma lies an unsettling revelation: consciousness is not an impartial observer but a catalytic force, shaping the contours of reality with each act of perception. The observer effect, that cornerstone of quantum mechanics, suggests a reality so intimately entangled with awareness that to observe is to create. In this delicate interplay, the boundary between self and cosmos dissolves, leaving us not as outsiders but as co-creators of existence itself.

Imagine each thought, each flicker of attention as a brushstroke on an infinite canvas. The quantum field is the palette, a reservoir of boundless hues and textures, waiting to be animated by the intent of the beholder. This interplay transforms the observer into both artist and art —a dance of mutual becoming where every choice ripples through the cosmic ether, leaving indelible marks on the fabric of being.

The Infinite Tableau of Possibility

The quantum field is a sea of infinite "perhaps," a boundless expanse where every possibility exists in superposition, awaiting the gentle caress of consciousness to be birthed into actuality. In this infinite tableau, we glimpse the ancient hermetic wisdom: "As above, so below." Each choice, each flicker of will, sends ripples cascading across the vast expanse, shaping worlds yet unborn.

This understanding is not merely theoretical; it is deeply transformative. If reality itself bends to the gaze of consciousness, then our inner worlds—our fears, dreams, and desires—are not confined within us but extend outward, sculpting the contours of existence. The quantum veil reveals not a deterministic universe but a coalescence of freedom and creativity, where every act of perception is an act of creation.

Science as Sacred Exploration

Here, at the edges of the quantum realm, science and mysticism entwine in an intricate embrace. The language of equations becomes indistinguishable from the poetry of the mystic; the quest for empirical truth mirrors the spiritual journey toward transcendence. To probe the quantum field is to stand on sacred ground, where the known and the unknowable converge, each illuminating the other.

This convergence does not diminish the rigor of inquiry but elevates it, urging us to approach the mysteries of existence with both precision and reverence. The quantum veil beckons us to a sacred science, one that recognizes the cosmos not as an inert machine but as a living, breathing continuum—a divine mosaic of light and shadow, form and formlessness.

The Mystery Beyond the Veil

The veil is not a barrier but an invitation, a luminous threshold that dares us to step beyond the confines of certainty. To peer into its depths is to encounter the infinite, to feel the gravitational pull of the unknown drawing us toward a deeper truth. And yet, the closer we approach, the more the mystery deepens, whispering that the ultimate revelation lies not in the answers but in the endless unfolding of the questions.

In this dance with the ineffable, we find not resolution but wonder—a quiet awe that humbles and exalts in equal measure. The quantum veil becomes a mirror, reflecting back not only the cosmos but our place within it, reminding us that to seek is itself an act of creation, a weaving of the tapestry that binds us to the stars.

Toward the Infinite

And so, we begin—not with certainty, but with curiosity; not with conclusions, but with an invitation to explore. The quantum field, the shimmering veil, is both map and territory, offering glimpses of a reality so vast, so intricate, that it defies the confines of language and thought. It calls us to embrace the paradoxes, to revel in the mysteries, and to recognize that the path forward is not a line but a spiral, an endless unfolding toward the infinite.

Through the veil, we move—not to conquer but to commune, not to master but to marvel, as seekers of a truth that will forever elude our grasp and yet forever fill our hearts with wonder.

CHAPTER 2: LIGHT AND SHADOW – THE DANCE OF DUALITY

Before time stretched its relentless thread and before space unfurled its vast canvas, there was only the pulse of pure potential—a boundless void, shimmering with the promise of creation. This was not emptiness, but a fertile silence, pregnant with opposites waiting to take form. Within this primal stillness, the first whispers of duality stirred: light and shadow, presence and absence, fullness and void. These twin forces would soon emerge, weaving the cosmos into a harmonious symphony of contrasts.

The dance of duality is not merely the story of creation; it is the essence of existence itself. To understand it is to peer into the heart of reality and witness the eternal interplay between forces that seem opposed, yet are inseparably intertwined.

Light's Paradoxical Nature – A Beacon of Mystery

Light, the harbinger of visibility and life, carries within it a profound enigma. Is it a wave, flowing seamlessly through the fabric of space, or is it a particle, discrete and countable, bounding from one point to another? Quantum mechanics answers with a riddle: light is both, and it is neither, depending on how it is observed. This duality defies logic, transcending the limits of classical reasoning. It forces us to abandon the comfort of absolutes and embrace a more fluid understanding of truth.

This dual nature of light finds its echo in the wisdom of the ancients. The yin-yang symbol from Taoist philosophy illustrates a similar paradox: two forces, darkness and light, each containing the seed of the other, perpetually in flux. Like light's duality, yin and yang are not opposites to be reconciled but complements that define each other. Just as light's identity shifts with the observer's gaze, so does reality ebb and flow in a ceaseless dance of transformation.

In this revelation lies a profound lesson: the essence of light, and indeed of all existence, is not fixed. It is a process, an unfolding narrative shaped by perception. To understand light's paradox is to begin to grasp the fluid, dynamic nature of the universe itself.

The Shadow – A Silent Architect of Form

While light dazzles and reveals, shadow lies quietly in its wake, often misunderstood as its antithesis. Yet shadow is not the absence of light but its silent collaborator. It is within the shadow that the contours of form emerge, the subtleties of texture are revealed, and the depth of reality is understood. Without shadow, light would be unbounded, formless—a radiance without definition.

In the spiritual realm, the shadow is equally profound. Mystics speak of the "shadow self," the unseen and unacknowledged parts of our being that hold the keys to transformation. Just as light cannot exist without shadow, the soul cannot achieve wholeness without embracing its darker aspects. The shadow is fertile ground, a reservoir of potential waiting to be illuminated.

In the quantum domain, this relationship manifests as a delicate interplay between matter and antimatter, presence and absence. Particles flicker into existence from the void, only to vanish again, suggesting that all of reality is born from this oscillation between the seen and the unseen. Shadow, then, is not a void but a crucible—a space where form and meaning are forged.

Creation and Destruction – The Cosmic Pulse

At the heart of the cosmos is a rhythm, a ceaseless pulse that beats across scales from the subatomic to the galactic. This is the rhythm of creation and destruction, the building and unbuilding of worlds. Stars are born in fiery explosions, only to collapse into the silent embrace of black holes. Particles arise from the quantum vacuum, only to dissolve back into its depths. Nothing in the universe is static; everything exists in a state of becoming and unbecoming, a cycle as ancient as time itself.

This rhythm is not chaotic but harmonious, a balance that sustains the cosmos. The mystics recognize this pulse as the breath of the Divine, an eternal inhalation and exhalation that gives rise to all things. The physicist sees it in the quantum field, where particles and waves flicker and oscillate in a dance of impermanence. It is a reminder that creation and destruction are not opposites but partners, each necessary for the other's existence.

Ancient cultures have long intuited this truth. They saw it in the cycles of the seasons, the waxing and waning of the moon, the ebb and flow of tides. In every death, there is the seed of rebirth; in every ending, a new beginning. This cycle is not a tragedy but a testament to the vitality of duality, a celebration of the interplay that gives life its vibrancy.

The Observer's Role – A Mirror of Consciousness

Quantum mechanics introduces a startling idea: the observer is not a passive witness but an active participant in the unfolding of reality. To observe is to collapse the wave of potential into the solidity of form. The act of looking shapes what is seen, intertwining the observer with the observed in a relationship so intimate that it dissolves the boundary between them.

This insight mirrors the mystic's understanding of the self. The journey inward is a dialogue with light and shadow, an exploration of the opposites within. Thoughts arise and dissolve like quantum particles, shaped by the awareness that observes them. In both realms, reality is revealed as relational, a dynamic interplay where the knower and the known are inextricably linked.

This intertwining challenges our notions of separation. The observer is not external to the observed; they are co-creators of a shared reality. In this interplay, the universe becomes a mirror, reflecting not only its mysteries but our own.

The Balance of Yin and Yang – Harmony in Opposition

The concept of yin and yang offers a profound understanding of balance. It teaches that opposites are not contradictions but complements, each necessary for the other's existence. Light defines darkness, and darkness gives meaning to light. This dynamic equilibrium is echoed in the quantum world, where particles and antiparticles, forces of attraction and repulsion, maintain the delicate harmony of the cosmos.

To embrace yin and yang is to see the beauty in balance, the wisdom in duality. It is to recognize that life's contrasts—joy and sorrow, gain and loss—are not adversaries but partners in a larger dance. The quantum field reveals this truth in its symmetry, showing us that opposites are not separate but interconnected, bound by an underlying unity.

The Illusion of Separation – Entanglement's Revelation

Quantum entanglement challenges the very notion of separateness. When particles become entangled, their states remain intertwined, regardless of distance. This phenomenon suggests that beneath the surface of multiplicity lies a seamless unity, a web that connects all things.

Mystics have long taught this truth, that the boundaries we perceive—between self and other, here and there—are illusions. Entanglement offers a scientific parallel to this spiritual insight, revealing that duality is not division but diversity, a manifestation of unity in its myriad forms.

Reality as a Sacred Dance

Reality is not a fixed state but a dynamic process, a sacred dance of opposites that gives rise to form and meaning. In the quantum realm, particles behave not as static entities but as waves of possibility, becoming real only when observed. This reflects the mystic's vision of the world as Maya, a play of form and formlessness where the Divine expresses itself in infinite variation.

To see reality as a dance is to surrender to its mystery, to embrace the fluidity of truth. Light and shadow, creation and destruction, self and other—all are partners in this eternal movement. This is the essence of duality: the recognition that opposites are not contradictions but complements, each holding the essence of the other, each necessary for the whole.

The Gift of Duality – A Path to Unity

Duality is not an obstacle but a gift, a lens through which we glimpse the infinite complexity of existence. It reminds us that unity is not the absence of difference but the harmonious interplay of opposites. In the dance of light and shadow, we find not only the structure of the cosmos but the reflection of our own souls—a reminder that we, too, are threads in this eternal tapestry, woven into the luminous dance of being.

CHAPTER 3: ENTANGLED SOULS – THE MYSTERY OF QUANTUM ENTANGLEMENT

In the depths of existence, where the visible dissolves into the imperceptible, an invisible bond whispers of connection—a bond that does not heed the confines of space nor the ticking of time. It stretches like an unseen symphony, linking the infinitesimal to the infinite, binding particles, stars, and souls in a shared resonance. This is the essence of quantum entanglement, a phenomenon that transcends the boundaries of matter to suggest an intimacy between all things—a quiet yet unshakable union that persists regardless of separation.

Through this enigma, a truth unfurls: existence itself is relational, a web of interdependence where no fragment can truly be isolated. To understand entanglement is to glimpse a world where each life, each moment, ripples into the next, resonating across an unseen matrix of unity.

The Bond That Defies Distance

Quantum entanglement offers a narrative of connection that defies our earthly logic. When two particles become entangled, their states become so deeply intertwined that the action of one instantaneously affects the other, no matter the gulf between them. This linkage, immune to the constraints of distance, defies the classical notion of causality, whispering of a deeper reality where space itself is no obstacle.

The mystics have long intuited such a bond. To them, the soul is not an isolated flame but a spark in a cosmic conflagration, inseparable from the fire that births it. Ancient traditions spoke of threads of fate, invisible cords tethering lives together, echoing the quantum truth that connection is the essence of being. In this view, individuality is not erased but transformed—each being is a node in a vast network, their actions rippling across the collective whole.

Entanglement as Sacred Design

To view quantum entanglement solely as a physical phenomenon is to glimpse only the surface of its profundity. Beneath the scientific explanation lies a sacred design—a hidden geometry of interconnection that links all things. Imagine a web where each soul is a luminous point, every connection forming an intricate, ever-shifting pattern of light. This design, invisible yet enduring, reveals that every thought, every act of will, reverberates across existence, touching countless lives in ways unseen.

This sacred architecture is not a metaphor but a lived reality. Empathy, love, and shared awareness hint at the interconnectedness entanglement describes. The mystic sees this geometry in moments of profound unity—in the embrace of another, in the unspoken understanding between strangers, in the silent communion with the universe. It is here, in the weaving of the visible and invisible, that we encounter the divine essence that binds all things.

The Collapse of Time – The Eternal Present

In the quantum world, entanglement challenges the tyranny of time. The instantaneity of communication between entangled particles suggests a realm where the linear progression of past, present, and future dissolves into an eternal now. This notion aligns with the mystic's experience of timelessness, where moments merge into a singular presence, a boundless "now" that holds all potential within it.

For the mystic in meditation, time ceases to be a river and becomes an ocean—limitless, unbounded, and whole. In this state, the divisions of before and after fall away, leaving only the recognition of unity. Quantum entanglement hints at a similar timelessness, suggesting that connection is not bound by temporal limits but exists outside the frameworks we impose upon it. In this eternal present, all is one, and all is here.

The Hidden Song of the Universe

Entanglement reveals an underlying harmony that sings through all existence, a resonance so subtle it often escapes the senses. It is a melody woven from silence, a vibration that links particles and people, a whisper of interconnectedness that hums through the cosmos. To attune oneself to this song is to feel the pulse of creation, a rhythm that binds all things in unity.

The mystics have described this resonance as the voice of the Divine, the breath of the infinite moving through all that is. It is a reminder that separation is an illusion, that beneath the surface of individuality lies an unbroken continuum. This voice invites us to listen—not with the ears, but with the soul—to hear the music of existence and recognize ourselves as both singer and song.

The Geometry of Unity

Imagine existence as a crystalline lattice, each soul a point of light in a structure of infinite complexity. Quantum entanglement hints at such a geometry, where each connection between particles forms an unseen pattern of resonance. These connections transcend the physical, creating a matrix of relationships that stretches beyond what the senses can perceive.

The mystic's vision aligns with this understanding. They speak of sacred geometries—mandalas, patterns, and forms that reflect the unity of the cosmos. Each relationship, each bond, is a line in this vast design, a thread that links one soul to another in a web of meaning.

To view existence through this lens is to see every encounter as sacred, every interaction as part of a divine plan that unfolds endlessly.

Reflection and Reciprocity

Entanglement is not merely connection; it is reflection, a mirror held up to the universe itself. To be entangled is to exist in a state of reciprocity, where the act of observing or acting upon one part instantly affects another. In this mirroring lies a profound truth: that every soul reflects the whole, that each action is both an expression of individuality and a ripple in the collective.

For the mystic, this reciprocity is a reminder of the divine within all beings. To see another is to see oneself, to recognize that each life reflects the infinite. This mirroring dissolves the barriers of selfhood, revealing that the boundaries between "I" and "you" are as illusory as the separation between stars in the night sky. Entanglement shows us that we are not solitary but symphonic, each note resonating with all others.

The Illusion of Separation

One of the deepest lessons of entanglement is the illusion of isolation. Particles, though separated by vast distances, remain fundamentally connected, their fates intertwined. This phenomenon speaks to a reality where separateness is a construct, a surface-level perception that masks the profound unity beneath.

The mystic has long understood this truth. They see the world not as fragments but as facets of a single jewel, each reflecting the same light. In this understanding, harm to another is harm to oneself; love for another is love for the whole. To live with this awareness is to transcend the illusion of division, to recognize that all beings are part of the same vast, interconnected whole.

Entanglement as a Path to Compassion

If entanglement teaches us anything, it is that our actions, thoughts, and intentions do not exist in isolation. Each choice sends ripples through the web of being, touching lives in ways we may never see. This awareness calls us to compassion, to a recognition that the well-being of one is tied to the well-being of all.

The mystic understands this as the essence of love —a force that binds, uplifts, and heals. Quantum entanglement offers a scientific echo of this truth, showing that connection is not optional but inherent, that we are, by nature, relational beings. To live with this awareness is to act with reverence, to honour the sacred interconnectedness that defines existence.

The Infinite Unity

Quantum entanglement unveils a universe where all things are bound in an intricate web of relations, a cosmos where separation is but a fleeting illusion. This unity is not an abstraction; it is the fundamental nature of reality, a truth that resonates in both the equations of physics and the insights of mysticism.

In this unity, we find solace. We are not alone, not adrift in an indifferent universe, but part of a vast and loving whole. Entanglement reveals that every soul, every particle, every star is connected, held in an embrace that stretches beyond time and space. This is the ultimate mystery, the profound beauty of existence: that we are, and have always been, one.

CHAPTER 4: THE OBSERVER'S GAZE – PERCEPTION AS REALITY

In the deep and silent chambers of the cosmos, where form and formlessness intermingle, there is a gaze—a gaze that shapes reality, that brings worlds into being. This is the gaze of the observer, an ancient, primordial awareness that watches and, in watching, moulds existence. Quantum mechanics reveals that mere observation can collapse a wave of possibility into solid fact, as if reality itself yearns to be seen, to emerge from the shadows of potential under the watchful eye of consciousness. And yet, this mystery transcends physics, touching upon the very soul of mysticism, where the act of perceiving is a sacred rite, and the mind, like a mirror, reflects the infinite.

The Paradox of the Observer Effect
In the quantum realm, where particles shimmer in a sea of possibilities, observation acts as a catalyst, collapsing waves of potential into definitive states. The observer effect defies our understanding of objectivity; it suggests

that reality itself is participatory, that the act of seeing is not passive but creative. The mystic, too, has always known this truth—that perception is not mere reception but creation, that consciousness is an artist painting reality with each thought, each glance, each moment of awareness.

What does it mean, then, to observe? To observe is to shape, to sculpt reality with the force of attention, to choose among infinite possibilities the one that will become. In this choice lies a paradox: to see is to limit, to fix something fleeting into form. And yet, within this limitation lies liberation, for it is only in being observed that reality emerges from the void, that the world, in all its beauty and complexity, takes shape.

The Mind as Mirror and Sculptor

Within each of us lies a mirror—a mind capable of reflecting the vastness of existence, a consciousness that does not merely record but interprets, that sees and, in seeing, gives meaning. This mirror is not passive; it is a sculptor's hand, an alchemist's tool, transforming the raw material of perception into the structure of reality. In quantum mechanics, the observer effect tells us that particles behave differently when observed, as if responding to the gaze of consciousness, altering their course in reverence to the act of being seen.

The mystic understands this mirror as both a window and a veil, a pathway to the divine and a barrier of perception. In meditation, the mystic polishes the mirror of the mind, seeking to see reality without distortion, to look upon the world as it is, unclouded by thought and bias. For in seeing clearly, the mystic aligns with the very act of creation, merging with the cosmic gaze that underlies all perception, all awareness. To observe, then, is not simply to watch—it is to participate in the unfolding of existence, to bring forth reality from the depths of potential into the light of form.

The Cosmic Witness

Behind every gaze, behind every act of perception, lies the question: is there a cosmic observer, a great witness who sees all things? Mysticism speaks of a Divine Presence, an infinite consciousness that holds all reality within its embrace, that sees without judgment, without attachment. In this cosmic witness, the mystic sees the ultimate source, the origin of all awareness, the eye that never closes, the mind that never rests.

Quantum mechanics does not speak of gods, yet it hints at a unity of observer and observed, an interconnectedness that suggests the universe itself is aware. Could it be that the act of observing is, in essence, a participation in the mind of the cosmos, that to see is to touch the divine, to align oneself with the vision of the infinite? Perhaps, in every gaze, in every act of perception, we are echoes of a greater seeing, reflections of a consciousness that holds all things within itself, that witnesses not from without but from within, as part of the whole.

Perception as the Alchemy of Reality

If observation shapes reality, then perception becomes a sacred alchemy, a transformative process by which the raw potential of existence is transmuted into the substance of our lives. Each moment of awareness is a choice, a decision that echoes across the quantum field, creating pathways of possibility, shaping the contours of experience. The mystic understands this alchemy as intention, the power of mind and heart to influence the unfolding of reality, to bring into form that which lies hidden in the depths of the soul.

Just as the alchemist transforms lead into gold, so too does consciousness transform the unmanifest into the manifest, turning possibility into experience, dreams into reality. This alchemy is subtle, invisible, yet its effects are profound, for in every thought, every glance, every moment of awareness, we participate in the creation of the world. To perceive is to be an artist of existence, to wield the power of mind as a brush, painting reality in strokes of intention and attention.

The Eye of the Soul
In the depths of the soul lies an eye, a silent witness that watches without judgment, without attachment. This inner eye is not bound by the limitations of the senses; it sees beyond appearance, beyond form, perceiving the essence that lies within. The mystic calls this the eye of the soul, an inner vision that transcends the physical, that sees reality as it truly is—whole, interconnected, boundless.

To open the eye of the soul is to awaken to a higher perception, to see not merely with the mind but with the heart, to understand that each thing, each being, reflects the same divine light. In this vision, duality dissolves, and the world is revealed as a unity, a single presence manifesting in infinite forms. This is the ultimate act of perception, the awakening of consciousness to itself, the realization that the observer and the observed are one.

The Universe as a Living Mirror
In the silence of deep contemplation, the mystic and the

physicist alike come to see the universe not as a collection of objects but as a living mirror, a vast field of awareness reflecting itself in countless forms. Every particle, every star, every soul, is a facet of this mirror, each reflecting the whole, each containing within itself the essence of all that is. To observe is to look into this mirror, to see oneself reflected in the vastness of existence, to recognize that each act of perception is a communion, a merging of self with the cosmos.

This living mirror reveals that reality is relational, that nothing exists in isolation, that each being is connected to every other through the simple act of seeing, of being seen. In this realization, the boundary between self and other begins to blur, and we come to understand that to observe the world is to observe oneself, that to know another is to know the divine. The universe, then, is not a machine but a soul, a vast consciousness reflecting itself in the infinite gaze of the observer.

The Sacred Act of Seeing

To see is to create, to participate in the divine unfolding of existence. The mystic approaches this act of seeing with reverence, understanding that each moment of awareness is a sacred ritual, an opportunity to touch the infinite. In quantum mechanics, the observer effect reveals this truth in scientific terms, showing that the act of observation is not passive but powerful, that consciousness itself is a force that shapes reality.

This sacred act of seeing calls us to mindfulness, to an awareness that goes beyond thought and feeling, that touches the essence of being. It invites us to look upon the world with fresh eyes, to see each thing as sacred, each moment as holy, each breath as a gift. In this vision, life becomes not a series of random events but a symphony of perception, a dance of awareness where each gaze, each thought, is a note in the song of creation.

The Dance of the Observer and the Observed

The mystic speaks of a dance, a movement of spirit and form, where the observer and the observed meet in a space beyond separation. Quantum mechanics echoes this dance, revealing that reality is a process, a dynamic interplay where consciousness and matter are entwined, where each defines and shapes the other. In this dance, we see the beauty of duality, the harmony of opposites, the unity that lies within diversity.

To observe is to dance with the world, to engage in a relationship that transcends the self, to become part of the unfolding of existence. In this dance, the observer is not separate from the observed but is a partner in the creation of reality, a co-creator in the cosmic play. This is the mystery of perception, the recognition that we are both the dancers and the dance, that reality is a living process, a continual act of creation in which we are both participants and witnesses.

The Mystery of the Divine Gaze

And so, we return to the mystery of the observer's gaze, to the recognition that to perceive is to touch the divine, to participate in a reality that is both seen and unseen, both known and unknowable. Is there a cosmic observer, a consciousness that holds all things within its sight? The mystic answers with a quiet certainty: yes, there is an eye that sees beyond form, beyond time, beyond the limitations of perception. It is an eye that looks through us, that sees through us, that knows us in ways deeper than we know ourselves.

To align with this divine gaze is to awaken to our own nature as observers, as beings of awareness capable of seeing beyond the surface, capable of touching the essence. It is to understand that we are part of the cosmos, that our awareness is a fragment of the infinite consciousness that watches, that loves, that creates. In this understanding lies peace, a peace born from knowing that we are seen, that we are known, that we are part of a greater whole.

The Infinite Journey of Perception
The journey of perception is an infinite one, a path that leads us ever deeper into the mystery of being. To see is to journey, to explore the layers of reality, to uncover the hidden depths of existence. Each moment of observation is a step along this path, a glimpse into the nature of the self, the world, the divine. In the end, perception is not an act but a process, a continual unfolding that mirrors the universe itself, an endless revelation of the infinite within the finite.

And so, we continue to look, to see, to be seen, each of us a fragment of the cosmic gaze, each of us a witness to the beauty and mystery of life. In this journey, we find the truth of existence—not as a fixed reality but as a flowing, living awareness, a boundless consciousness that moves through us, that sees through us, that loves through us. This is the gift of perception, the sacred act of seeing, the eternal dance of the observer and the observed.

CHAPTER 5: THE VOID AND THE PLENUM – EMBRACING EMPTINESS

In the fathomless expanse where stars are forged and galaxies pirouette through eternity, there resides a paradox too immense for the senses to grasp. What appears as absence is brimming with latent energy, and what seems to be silence hums with the pulse of creation. The Void, a concept that has both haunted and inspired physicists and mystics alike, is no barren nothingness but an infinite wellspring of potential, the origin and eventual resting place of all that exists. It is the unspoken source of life, a fullness so vast it wears the guise of emptiness.

To contemplate the Void is to venture beyond the limitations of perception, to peer into the depths of a mystery that cannot be named. It is an invitation to dissolve the boundaries of self and discover a truth that eludes the grasp of language—a truth that dances at the edge of awareness, whispering of infinite possibility.

The Quantum Vacuum: A Cauldron of Becoming

In the lexicon of physics, the quantum vacuum is often misunderstood as mere emptiness, a void devoid of substance. Yet this "emptiness" is anything but barren. It is a teeming sea of possibility, an unseen matrix from which particles emerge and into which they vanish in an endless, ceaseless rhythm. It is a plenum, a hidden fullness vibrating with potential energy, the silent architect of the material world.

This vacuum is the stage upon which the universe performs its grand drama, its particles and waves flickering into existence like ephemeral actors. Beneath the apparent stillness lies a ceaseless hum—a symphony of energies too subtle for the senses but foundational to all that is. The physicist sees this as a field alive with fluctuations, a substrate from which reality itself is sculpted. The mystic, in turn, beholds the same essence as the Divine Void—a sacred emptiness pregnant with the promise of creation.

To glimpse this quantum cauldron is to encounter the paradox of fullness within emptiness. It is the ultimate reminder that what appears absent is, in truth, the fertile ground for all becoming.

The Unheard Symphony and the Unseen Dance

To speak of emptiness is to confront the paradox of silence—the silence beneath sound, the stillness that gives birth to motion. Just as the vacuum underlies all energy, so does the Void underpin every experience, holding within it the seeds of sound, colour, and form. This silence is not mute; it is a deep reservoir of meaning, a foundation upon which the cacophony of existence rests.

For the mystic, entering the Void is not an act of retreat but one of profound engagement. In the stillness, they touch the unstruck sound, the note that resonates before vibration. This is the Nada, the primal hum, a frequency that echoes in the chambers of the soul, reminding us that the emptiness we fear is, in truth, the source of all we hold dear.

Zero-Point Energy: The Eternal Pulse

Within the quantum realm, even the so-called void is animated by zero-point energy—a baseline resonance that persists even in the absence of matter. This is the unquenchable heartbeat of the cosmos, a silent rhythm that endures through the stillness, a pulse that binds the infinite to the finite.

For those who seek the divine in silence, zero-point energy becomes a metaphor for the eternal—a presence that vibrates beneath the veil of perception. To physicists, it is an anomaly, a challenge to the notion of true emptiness. To the mystic, it is a signpost pointing to the sacred: a reminder that the absence we perceive is merely the backdrop for presence in its most subtle form.

In contemplating zero-point energy, one begins to grasp the truth of the Void: that it is not a barren wasteland but a reservoir of infinite potential, a reminder that creation begins not in chaos but in the profound stillness of being.

The Womb of Creation: Emptiness as Origin

The Void is more than a field of possibility; it is the cosmic womb, the sacred vessel where form is conceived and nurtured before taking its first breath. Within this womb, stars ignite, galaxies are spun, and life is whispered into existence. It is a space of infinite patience, where time itself waits for the moment of becoming.

In the quantum vacuum, particles emerge spontaneously, birthed from the boundless energy that pervades emptiness. Mystics see the same dynamic in the Divine Void, where all creation arises from a space of stillness, a silence so profound it gives rise to sound. In both realms, emptiness is not a negation but a promise—a reminder that within nothingness lies everything, that in the absence of form resides the potential for infinite forms.

To enter this womb is to surrender to a vast humility, to recognize oneself as both participant and creation, both the sculptor and the clay. It is to acknowledge that existence itself is an act of grace, an unfolding that begins and ends in the arms of the Void.

The Paradox of Fullness and Emptiness

The Void challenges the mind's desire for clarity, confronting us with a truth that is both simple and incomprehensible: emptiness is fullness, and fullness is emptiness. In the quantum world, this is evident in the vacuum, a field that appears devoid yet teems with energy. In mysticism, it is the Void that cradles all existence, a space so vast that it cannot be grasped, so profound that it must be felt.

To live in alignment with this paradox is to release the need for certainty. It is to dwell in the mystery, to recognize that life's essence cannot be captured in words or concepts. It is to see that silence is the foundation of every sound, that stillness is the ground upon which every movement unfolds. This is the wisdom of the mystic, the insight of the physicist—a recognition that emptiness is not an absence but the essence of presence itself.

The Dance of Creation and Dissolution

The Void is not static; it is dynamic, a rhythm of creation and dissolution that mirrors the breath of the universe. Particles emerge from its depths, linger briefly in existence, and dissolve back into its embrace. Stars are born, blaze brightly, and collapse into silence. Life arises, grows, and returns to the soil from which it sprang. This is the dance of the cosmos, a rhythm that speaks of impermanence and renewal.

To live in harmony with this dance is to embrace the transient nature of existence. It is to see that every ending is a beginning, that every loss is a transformation, that every emptiness is an invitation to fullness. The Void teaches us to let go, to trust in the cycles of creation, to recognize that in the dissolution of form lies the promise of new life.

The Presence Beneath Perception

In the heart of the Void lies a presence so subtle, so all-encompassing, that it eludes the grasp of the senses. It is not a being but a becoming, not a force but a field, an awareness that permeates all things. Mystics speak of this presence as the Divine, the source that is both within and beyond, immanent yet transcendent, the silent witness to all that unfolds.

Physicists encounter this presence in the quantum vacuum, the unchanging field that underlies the flux of particles and waves. It is a reminder that beneath the chaos of existence lies a stillness, a ground of being that endures. To touch this presence is to encounter the sacred, to feel the infinite within the finite, to recognize that the essence of reality is not in its forms but in the formless.

The Gift of the Void: Freedom and Possibility

The Void offers a gift that is both humbling and liberating: the recognition that life is not a fixed reality but an infinite field of possibility. In the quantum vacuum, we see a world in flux, a cosmos in constant motion, where nothing is static and everything is becoming. In the mystic's journey, we find the same truth —a call to release attachment, to embrace the unknown, to dwell in the openness of being.

This freedom is not a negation but an affirmation. It is the understanding that emptiness is not a void to be filled but a space to be explored, a canvas upon which the infinite paints itself. To embrace the Void is to embrace life in its fullness, to see that within each moment lies the potential for transformation, that within each breath resides the infinite.

Living with the Void: A Path of Reverence

To live with an awareness of the Void is to walk a path of reverence. It is to see that every sound arises from silence, that every form is shaped by emptiness, that every life is held within the boundless embrace of the formless. It is to recognize that we are not separate from the Void but part of it, that our existence is an expression of its infinite creativity.

This is the wisdom of the Void: that to surrender to its silence is to find our voice, that to dwell in its stillness is to discover movement, that to embrace its emptiness is to uncover fullness. It is a call to live not in fear of the unknown but in wonder at the infinite possibilities it holds—a reminder that within the Void lies the essence of all that is.

CHAPTER 6: THE DANCE OF PROBABILITY – EMBRACING UNCERTAINTY

In the core of existence, where reality flutters on the edge of becoming, there is a dance—a delicate, ceaseless movement where certainty fades, and only probability remains. This is the dance of potential, a cosmic choreography woven from the threads of uncertainty, where every particle, every thought, every event exists as a possibility, a whisper of what could be. Quantum mechanics reveals this truth in the language of probabilities, teaching us that the universe does not operate on fixed certainties but on a field of potential, a realm where reality arises from the dance of chance.

To understand this dance is to embrace a new way of seeing, a perspective that accepts the fluidity of existence, the inherent unpredictability that underlies all things. Mysticism, too, speaks of uncertainty as a pathway to wisdom, a surrender to the unknown that leads not to chaos but to freedom. In this chapter, we explore the dance of probability as both a scientific and spiritual truth, a reminder that life is not a solid structure but a flowing river of possibilities.

The Quantum World: A Symphony of Chances

In the quantum realm, particles do not follow deterministic paths; they exist in a cloud of potentialities, a haze of possible locations, spins, and states, each awaiting the moment when observation will bring one of them into being. This probabilistic nature of the quantum world defies the classical notion of certainty, revealing instead a reality that is less about absolutes and more about tendencies, a place where events unfold according to the gentle nudge of probability rather than the rigid law of cause and effect.

To live within this quantum field is to live in a state of openness, to dwell within a spectrum of possibilities, each as real as the other, each a glimmer of what might come to be. The mystic recognizes this openness as the essence of creation, an embrace of potential that allows for the flowering of infinite forms. In this state, there is a beauty that arises from uncertainty, a dance that celebrates the unknown, a freedom that exists not in knowing but in trusting the flow of possibility.

The Mystery of the Wave Function: Reality as Potential

At the heart of quantum mechanics lies the wave function, a mathematical expression that encapsulates all possible states of a particle. The wave function is not a single reality but a field of probabilities, a description of potential rather than actuality. It is a dance of might-bes, a symphony of possible outcomes, a song that resonates with the mystery of existence. When we observe a particle, the wave function collapses, and one possibility becomes real, but until that moment, reality remains open, fluid, undefined.

The mystic understands this fluidity as the nature of existence itself, a dance of form and formlessness where nothing is fixed, where every moment holds the potential for transformation. In meditation, the mystic dwells within this openness, touching the unmanifest, feeling the pulse of potential that lies beneath the surface of reality. To live in alignment with the wave function is to live in a state of surrender, to release the need for control and to trust in the unfolding of life's probabilities, a trust that transcends fear and welcomes the mystery of being.

Uncertainty as the Gateway to Freedom
In embracing uncertainty, we discover a hidden freedom—a liberation from the confines of expectation, a release from the need to know. Quantum mechanics shows us that reality itself is uncertain, that particles behave unpredictably, that existence is woven from the fabric of chance. This uncertainty is not a flaw but a gift, an invitation to explore, to wonder, to approach life with curiosity rather than certainty.

Mysticism, too, teaches the wisdom of uncertainty, urging the seeker to let go of rigid beliefs, to open to the vastness of the unknown. In the mystical path, uncertainty is not a void but a doorway, a passage that leads to deeper understanding, to a state of mind that embraces possibility, that sees life as an adventure rather than a fixed journey. To dwell in uncertainty is to walk in freedom, to recognize that each moment is a chance, a choice, a step in the unfolding dance of probability.

The Cosmic Dice: Life as a Play of Chance

In the quantum realm, particles move as though guided by invisible dice, rolling their way through space and time, following paths that are determined not by certainty but by probability. Each event, each interaction, is a throw of the cosmic dice, a moment of chance that shapes the flow of reality. To see life in this way is to accept that not everything can be controlled, that existence is a game played on the edge of possibility, a dance where each step is unpredictable, each outcome unknown.

The mystic finds beauty in this play of chance, understanding that life's unpredictability is not a flaw but a form of divine play, a cosmic lila where the universe delights in the freedom of possibility. To surrender to this play is to embrace the truth that life cannot be pinned down, that each moment is fresh, new, alive with potential. This is the wisdom of chance, the understanding that uncertainty is a blessing, a reminder that we are participants in a dance that is ever-changing, ever-unfolding, a dance that holds the beauty of the unknown.

Probability as the Language of Creation

If the universe operates on probability, then creation itself is a process of choosing among possibilities, a selection of one reality from a field of potentials. Quantum mechanics shows us that each particle, each atom, is a decision, a choice made from an infinite sea of possibilities. In this way, probability becomes the language of creation, the voice that speaks existence into being, the whisper that calls forth the world from the void of potential.

The mystic understands creation in similar terms, seeing each moment as a choice, each thought as a seed, each action as a step in the unfolding of potential. To live as a creator is to embrace this dance of probability, to recognize that life is not a predetermined path but a journey of choices, a flow of decisions that shape the landscape of reality. This is the art of living, the skill of weaving one's own path through the field of possibilities, a path that honours the unknown, that celebrates the power of choice, that dances with the mystery of probability.

The Wisdom of Not-Knowing
In a universe governed by probability, certainty becomes a limitation, a boundary that closes the mind to possibility. The mystic, like the quantum physicist, learns to dwell in not-knowing, to accept that some truths lie beyond the reach of understanding, that reality is too vast, too complex, to be fully known. This wisdom of not-knowing is not ignorance but humility, a recognition that the mind is limited, that life is a mystery, that to know everything would be to close oneself off from the infinite.

To live in not-knowing is to embrace the fullness of existence, to approach life with an open heart, to trust that there is a greater intelligence at work, a pattern that we may not see but that guides the dance of probability. In this state, there is a peace that arises, a freedom from the burden of certainty, a joy that comes from accepting the unknown. This is the wisdom of the mystic, the freedom of the seeker, the peace that comes from dancing with life's uncertainties.

The Dance of the Known and the Unknown

In the quantum field, there is a constant interplay between the known and the unknown, a dance where particles flicker in and out of existence, where possibilities shift and change, where reality remains fluid, open, alive. This dance is not chaotic but harmonious, a rhythm that moves with a grace that transcends understanding. In the same way, life unfolds in a balance of known and unknown, a dance that invites us to step into the mystery, to trust in the beauty of what we cannot see.

The mystic finds joy in this dance, understanding that life is a flow, a movement that cannot be grasped but only experienced, a river that carries us not to a fixed destination but to a deeper understanding of the journey itself. To dance with the unknown is to live fully, to embrace each moment as a new beginning, to see each choice as a possibility, each experience as a chance to learn, to grow, to become.

The Divine Gambit: Creation Through Uncertainty
In the vastness of the cosmos, creation unfolds not as a deterministic plan but as a divine gambit, a play of chances, a dance of probabilities. This is not randomness but freedom, a freedom that allows for creativity, for spontaneity, for the beauty of surprise. In the act of creation, the universe plays with its own potential, exploring the infinite ways in which life can express itself, the endless forms that being can take.

The mystic sees this divine play in all things, understanding that life is not a fixed story but a poem, a song, a dance that is forever changing, forever new. To live in alignment with this play is to embrace the beauty of uncertainty, to see that each moment is a gift, each day a chance to dance with the unknown. This is the path of freedom, the way of joy, the wisdom that comes from knowing that life is a mystery, a game that invites us to play, to explore, to create.

The Infinite Possibilities of Existence

And so, we return to the dance of probability, to the recognition that life is a field of infinite possibilities, a realm where each choice, each thought, each action opens a new path, a new reality, a new way of being. In the quantum world, particles do not follow fixed paths; they explore, they wander, they choose among possibilities, creating reality with each moment of movement, each instant of interaction.

The mystic, too, walks this path of possibility, embracing life as an open field, a canvas where each moment is an opportunity to create, to discover, to become. In this dance, there is no end, no final destination, only the endless unfolding of potential, the infinite expression of the soul's journey. To embrace this dance is to live in alignment with the universe, to understand that life is not a fixed reality but a flow of becoming, a dance that invites us to move, to create, to love.

The Joy of Uncertainty

In the end, to embrace uncertainty is to embrace life itself, to understand that beauty lies not in certainty but in the freedom of the unknown. This is the gift of the quantum world, the wisdom of the mystic—a reminder that life is a dance, a celebration of possibility, a journey that holds the joy of discovery, the thrill of creation, the peace of surrender.

To live with joy in uncertainty is to live fully, to see

each moment as an invitation to dance, each day as an opportunity to explore, each breath as a reminder that life is a mystery, a mystery that does not ask to be solved but to be embraced. This is the dance of probability, the art of living with an open heart, the beauty of a life lived on the edge of possibility, where each step is a choice, each gaze a creation, each heartbeat a note in the song of existence.

CHAPTER 7: SUPERPOSITIONS OF SELF – EMBRACING PARADOX

In the quantum realm, particles inhabit an enigmatic liminality, existing in a state that defies the binary nature of existence. These particles, suspended in superposition, do not occupy a singular identity but inhabit a spectrum of potentialities—a delicate choreography of possibilities awaiting the moment of observation to crystallize into reality. In their unobserved state, they are neither here nor there but everywhere at once. This principle of quantum superposition offers a profound reflection of our own human condition, revealing the layered and fluid nature of selfhood.

To stand still in quiet introspection is to recognize within oneself the paradoxes that define us. We are beings of multiplicity, each of us carrying the echoes of countless "selves." There is the self we present to the world, the self we conceal, the self that dreams beyond boundaries, and the self-tethered by fears. Like particles in superposition, we exist as a myriad of overlapping identities, each one adding depth and complexity to our being. Embracing these paradoxes is not an act of resolution but one of acceptance—a recognition that our essence is not singular but dynamic, shifting and expanding as we traverse the unfolding journey of life.

The Paradox of the Many Selves

The idea of a singular, unchanging self is a comforting illusion, a structure we cling to in a world often marked by chaos. Yet, beneath this façade of singularity lies a more profound truth: we are beings of multiplicity, our identities woven from a rich tapestry of memories, aspirations, contradictions, and experiences. Quantum mechanics, with its principle of superposition, mirrors this multiplicity, showing that particles—and by extension, people—can embody opposites, contradictions, and even impossibilities all at once.

Within each of us resides a constellation of selves—the innocent child, the relentless seeker, the passionate lover, the fearless warrior, the compassionate healer. These facets of being are not fragments but integral parts of our whole, each one representing a unique perspective of our journey. The self is not static; it is a kaleidoscope of identities that shift and blend, forming new patterns with each turn of life's wheel. To embrace this multiplicity is to acknowledge that we are not confined to any one version of ourselves but are all of them, coexisting in a vibrant, ever-changing dance of being.

The Dance of Potentiality

In the quantum domain, particles in superposition exist in a state of pure potential—a field of possibilities where all outcomes coexist. It is only when observed that the particle chooses a state, collapsing into a single reality. This phenomenon serves as a potent metaphor for the human self, which also exists as a constellation of potentialities, waiting for the light of awareness to illuminate one path among many.

To live in alignment with this quantum truth is to understand that our identities are not preordained but are shaped by our choices and perceptions. We are beings of boundless potential, and each moment offers an opportunity to explore a new facet of ourselves. Life, like the quantum field, is not linear but multidimensional—a labyrinth of paths that intersect and diverge, creating a rich and textured narrative. The mystic recognizes this dance of potentiality as the essence of the soul's journey—a journey that has no fixed destination but is instead an endless unfolding of being.

The Mystery of Identity

In a society that prizes definition, the idea of a fluid and multifaceted self can feel unsettling. We are taught to craft identities, to anchor ourselves in narratives that provide a sense of solidity and purpose. Yet, in the sanctuary of silence, we sense that these narratives are incomplete. We are not bound by the constraints of a single name or story; we are more than the sum of our labels.

The mystic approaches identity as an ever-evolving mystery—a reflection of the infinite rather than a fixed construct. To embrace this mystery is to release the need for certainty and to allow the boundaries of selfhood to dissolve. Identity is not a static endpoint but a dynamic process, a journey that reveals new depths and dimensions with every step. In this process, we encounter a deeper truth: that the self is a paradox, both singular and multiple, defined yet infinite.

The Self as a Field of Possibilities

The quantum superposition of a particle represents a state of boundless potential, where all possibilities coexist without contradiction. Similarly, the self is not a singular entity but a field of potentialities, a vast and uncharted landscape where multiple identities and realities intersect. This understanding frees us from the constraints of linear thinking, inviting us to explore the infinite facets of our being.

The mystic sees the soul as a field of awareness that transcends ego and personality, a boundless presence unconfined by time or space. Within this field lies the freedom to embody different aspects of self, to move fluidly between identities, to experience life from a multitude of perspectives. The self, in its superposition, is a creative force—dynamic, expansive, and eternally becoming. It is not confined to the past or limited by the future but is alive in the present moment, rich with possibility.

Embracing the Contradictions of Being

To exist is to inhabit contradiction. We are beings of duality, embodying light and shadow, love and fear, wisdom and naivety. These opposing forces do not diminish us; they complete us, creating a harmony that transcends the sum of its parts. Like particles in superposition, we carry within us the capacity to embody opposites, to hold conflicting truths, to be both this and that.

In embracing these contradictions, we move closer to wholeness. The mystic understands that true unity is not the absence of conflict but the integration of opposites—a balance that allows for the full expression of being. To accept one's contradictions is to accept one's humanity, to recognize that every facet of the self has its place and purpose. In this acceptance, we find peace—not a peace born of resolution but one rooted in the harmony of complexity.

The Self as an Ever-Evolving Mystery

Like a particle in superposition, the self is a presence that defies definition, a being in constant transformation. In each moment, we are reborn, reshaped by experience, renewed by the currents of life. To live as an evolving mystery is to embrace the beauty of not-knowing, to see the self not as a fixed entity but as a process of becoming.

The mystic understands that the soul is not a puzzle to be solved but an infinite horizon to be explored. There are no final answers, no ultimate definitions—only deeper questions and richer complexities. In this unfolding, we discover that the self is not a destination but a journey, a continual expansion of awareness and understanding.

The Liberation of Embracing Paradox

Within paradox lies liberation. To embrace the superpositions of self is to let go of the need for certainty, to revel in the richness of complexity, to accept that we are both many and one. This liberation allows us to inhabit all that we are without fear or limitation, to dance between identities, to explore the fullness of our being.

The mystic sees this liberation as the essence of spiritual freedom—a journey into the heart of paradox that reveals the self as both finite and infinite, both bound and free. In this journey, we encounter a joy that arises not from answers but from openness, a peace that comes from accepting the multiplicity of the self, a love that flows from the recognition of our inherent wholeness.

The Divine Play of Self and No-Self

In the mystical tradition, the self is both a reality and an illusion—a construct of the mind that obscures the deeper truth of unity. Yet, even as an illusion, the self reflects the divine, a play of forms that hints at the formless. To embrace the self is to participate in the cosmic lila—the divine dance where identity is both real and unreal, fixed and fluid, present and absent.

The mystic understands that to lose the self is to find it anew, to see that identity is not a prison but a doorway to the infinite. In the interplay of self and no-self, we discover a joy that transcends boundaries, a peace that comes from surrendering to the mystery, a love that flows from the recognition that the self is, in its essence, a mirror of the divine.

The Infinite Self

The self, like a particle in superposition, is not bound by singularity but expands infinitely. With each moment, we are renewed, redefined, and reimagined. To live as an infinite self is to embrace the mystery of being, to see that identity is not a fixed point but an ever-unfolding journey.

This is the gift of superposition: the knowledge that we are not limited by any one version of ourselves but are beings of infinite potential. Each moment invites us to explore, to create, to become. In this journey, there are no conclusions, only endless beginnings—a continual dance of becoming that celebrates the paradox of existence and the boundless beauty of the self.

CHAPTER 8: TIME UNBOUND – BEYOND LINEAR PERCEPTION

Time, the enigmatic axis around which our lives revolve, appears to us as a flowing stream, relentlessly moving from what was to what will be. We measure it, track it, and bind our existence to its rhythm, yet beneath this orderly façade lies a profound mystery. The quantum world hints at a reality where time is not a linear thread but a boundless field, an intricate interplay of moments unanchored by direction. In this vision, the past and future are not separate shores but part of an unbroken whole, folded into the eternal present. For the mystic, time is not merely an abstraction but a veil that conceals the infinite—a mirage obscuring the timeless essence of existence.

To step beyond the illusion of linear time is to see the universe anew, to witness the fluidity of moments as they ripple through the cosmos, interweaving in ways that defy comprehension. This journey calls us to question the constructs we hold dear, to release our grip on the ticking clock, and to open ourselves to the boundless expanse of the eternal now.

The Mirage of Linearity

The human mind craves order, and so we divide time into measurable fragments: seconds, hours, years. These units, though practical, are fabrications—a scaffolding erected to make sense of the ceaseless flow of existence. Yet, the quantum realm shatters this illusion. Here, particles transcend our rigid notions of before and after, interacting across temporal divides as though past and future were merely different aspects of the same whole. Cause and effect blur; the arrow of time bends back upon itself, revealing a universe where time is not a river but a shimmering, multidimensional field.

In the stillness of contemplation, the mystic perceives this same timelessness. When the mind's chatter subsides, the heart opens to a vast, unbroken presence, a now that is neither fleeting nor fixed but eternal. This moment holds all moments within it, a singularity through which the past whispers its wisdom and the future hums with possibility. To glimpse this timeless state is to see that what we call the present is not a point on a line but a boundless expanse—a luminous portal into infinity.

The Quantum Continuum: A Field of Moments

Time, as understood through the lens of quantum mechanics, does not proceed in a straight line. Instead, it exists as a continuum—a boundless expanse where every moment interacts with every other, weaving an intricate web of connection. Particles, entangled not only in space but in time, defy the notion that the past is fixed or the future unwritten. They remind us that what was and what will be ever-present, folded into the intricate fabric of now.

For the mystic, this continuum is not merely a scientific insight but a spiritual reality. Time is not a march of moments but an eternal presence, a vast sea in which each wave touches every other, creating a harmony that transcends sequence. To embrace this vision is to step outside the confines of past and future, to dwell in the fullness of now, where the essence of all time converges in a singular, timeless presence.

The Paradox of Presence and Absence

If time is not linear, then the boundaries between past, present, and future dissolve, revealing a paradox where presence and absence coexist. The past does not vanish; it lingers as a living force, shaping the contours of now. Similarly, the future is not a distant horizon but an intimate presence, a potential that infuses each moment with its silent call.

In meditation, the mystic touches this paradox, feeling the echoes of what has been and the whispers of what might be, woven together in the stillness of now. To live in this awareness is to recognize that no moment is truly lost, that each breath carries within it the fullness of eternity. It is to understand that life is not a series of disconnected instants but a seamless whole, where every experience, every choice, reverberates across the continuum of being.

Cycles Beyond Sequence

Ancient wisdom traditions understood time not as a line but as a wheel, a sacred rhythm that turns and returns, mirroring the cycles of nature and the cosmos. The phases of the moon, the changing seasons, the rise and fall of civilizations—all speak to a reality where endings are beginnings, where life is not a journey from point A to point B but a spiral, an eternal recurrence that transcends the limits of linearity.

Quantum mechanics echoes this vision. Events may recur, particles may revisit past states, and the dance of creation and dissolution unfolds in patterns that defy conventional logic. This cyclical nature of time invites us to see life not as a series of irreversible steps but as a dynamic flow, a rhythm that pulses with renewal. In this view, every loss holds the promise of rebirth, every death the seed of new life. To live in harmony with this rhythm is to embrace the eternal return, to see each moment as both ancient and new, both ephemeral and eternal.

The Presence of Eternity in the Instant

In the quiet spaces of the soul, time dissolves, revealing a presence that is neither fleeting nor fixed but eternal. This is the mystic's encounter with the infinite—a state where each moment is complete in itself, where each breath holds the essence of all time. Eternity is not a far-off dream but a quality of the present, a depth that lies just beneath the surface of ordinary awareness.

Quantum mechanics hints at a similar depth within each moment. The quantum field is not bound by the linear progression of cause and effect but exists as an open dimension, a canvas where past, present, and future coexist as aspects of a unified whole. To touch this presence is to see beyond the clock's ticking, to feel the infinite within the finite, to recognize that time's true nature is not chronological but eternal.

The Timeless Self

If time is unbound, then the self, too, transcends its temporal constraints. We are not merely beings of memory and anticipation, bound by the stories of our past and the dreams of our future. We are timeless presences, souls that exist beyond the limitations of sequence, holding within us the fullness of all that was and all that will be.

The mystic understands the self as a consciousness that is both eternal and immediate, both infinite and intimate. This self is not confined to the narrative of linear time but reflects the timeless essence of being. To live as a timeless self is to release the burdens of regret and expectation, to see that who we are is not bound by what has happened or what might come but is always present, always whole.

The Interwoven Dance of Past, Present, and Future

If time is a continuum, then past, present, and future are not separate entities but interwoven aspects of a single reality. In the quantum realm, particles influence one another across time, blurring the boundaries between what was, what is, and what will be. This interconnection suggests that all moments are part of a larger dance, a symphony where each note resonates with every other.

The mystic sees this dance as a reflection of the soul's journey. Each experience, each moment, is a thread in the intricate web of being, a note in the melody of existence. To live with this awareness is to embrace the interconnectedness of all things, to see that each choice reverberates across time, that each step in the dance is both individual and universal, both finite and infinite.

Time as the Mirror of Consciousness

Time, as both quantum mechanics and mysticism suggest, is not an external reality but a reflection of consciousness, a construct shaped by the mind's need for order. The past and future are not fixed but fluid, shaped by perception, by the act of observation. Reality itself is not a linear progression but a co-creation, a dance between observer and observed.

For the mystic, this reflection is a mirror that reveals the depths of the soul, a field where each moment is an opportunity to touch the timeless, to see beyond the surface to the essence of being. In this mirror, time is not a prison but a spiral, a journey that leads not to an end but to a deeper understanding, a closer communion with the infinite.

The Eternal Now

The mystic's greatest insight is the recognition that all time converges in the present, that the now is not a fleeting instant but an eternal presence. This is the heart of timelessness—the understanding that each moment holds within it the essence of all moments, that life is not a series of disconnected events but a unified whole, alive with the presence of eternity.

To live in the eternal now is to see each moment as a doorway to infinity, to embrace the fullness of life as it unfolds, to recognize that the essence of time is not in its passage but in its presence. This is the freedom of a life unbound by sequence, the joy of a soul in communion with the infinite, the beauty of a journey that holds all moments within it, a journey that is not linear but eternal.

CHAPTER 9: COSMIC SYMPHONY – THE MUSIC OF THE SPHERES

In the uncharted expanse of the cosmos, where galaxies drift and stars ignite, a silent symphony unfolds— a melody too vast for the ear, too intricate for the mind, yet resonant in the soul. It is a music not born of instruments, but of vibration itself, a hum that pervades all things, from the infinitesimal dance of quantum strings to the grand revolutions of celestial spheres. String theory whispers of a universe woven from filaments of energy, each vibrating at its unique frequency, crafting the architecture of reality itself. This insight is not new but echoes an ancient belief, one held by philosophers and mystics: that existence is not chaos but composition, a harmonic interplay where every particle, every being, every moment vibrates as part of a unified song.

The ancients called this the music of the spheres. Pythagoras, that sage of geometry and harmony, saw in the heavens a divine resonance, an inaudible hymn that binds the cosmos in symmetry, proportion, and balance. To listen to this cosmic music is to align oneself with the rhythm of existence, to perceive a universe not as inert matter but as a living symphony, vibrating with beauty, coherence, and interconnectedness.

String Theory: Vibration as the Essence of Reality

In the smallest recesses of existence, at scales far beyond human perception, string theory posits that the building blocks of the universe are not particles but infinitesimal loops of energy. These strings vibrate, oscillating in multidimensional space, and their frequencies determine the nature of all that exists—matter, energy, even the forces that govern them. Each vibration is a note in the grand composition of reality, creating the diversity of forms and phenomena that we perceive.

To understand this is to glimpse a world where all things, from the solidity of stone to the ethereal dance of light, are expressions of the same fundamental resonance. Galaxies and atoms alike vibrate to the same cosmic rhythm, their differences mere variations on a universal theme. This vision dissolves the boundaries between the material and the immaterial, the visible and the invisible, revealing a universe unified not by substance but by sound.

The mystic has always intuited this truth. For the mystic, reality is not a collection of discrete objects but a single, interconnected whole, vibrating with the breath of the divine. The strings of string theory are, to the mystic, the threads of existence itself, each vibrating with intention, each contributing its unique note to the harmony of creation.

The Harmony of the Spheres: Pythagoras' Vision

Pythagoras, the philosopher-mystic, envisioned a universe governed by harmony, where the movements of celestial bodies created an inaudible music, a cosmic order expressed through proportion and balance. To him, numbers and ratios were sacred—they were not mere abstractions but the language of the divine, the blueprint of the cosmos itself.

This vision finds a modern parallel in string theory, where the vibrations of energy strings mirror the harmonic principles that Pythagoras revered. Each string vibrates in a manner akin to a musical note, creating patterns that resonate across dimensions, connecting the tangible with the transcendent. This harmony is not confined to stars and planets but flows through all things, from the smallest particle to the grandest galaxy.

To perceive the music of the spheres is to touch the sacred geometry of existence, to recognize that life itself is not random but rhythmical, a dance of resonance that flows through all things. It is to understand that the universe is not a machine but a melody—a living, breathing symphony.

Resonance: The Pulse of Existence

At the heart of this cosmic symphony lies resonance—the principle that vibration is the essence of being. Each string vibrates not in isolation but in harmony with the whole, creating a unity that transcends separation. This resonance is the thread that binds the universe together, connecting the infinitesimal with the infinite, the seen with the unseen.

The mystic experiences this resonance as a pulse that moves through all things, a rhythm that binds soul to soul, life to life, world to world. It is a reminder that the divisions we perceive are illusions, that beneath the surface of diversity lies a profound unity. To live in resonance with this cosmic music is to feel a peace that arises not from isolation but from connection, a sense of belonging that transcends the boundaries of time and space.

The Sacred Geometry of Sound

The ancients understood sound not merely as vibration but as a force that shapes and orders reality. Pythagoras taught that sound is a form of sacred geometry, a pattern that reflects the harmony of the cosmos. Modern science has affirmed this ancient wisdom through the study of cymatics, which reveals how sound creates intricate patterns and structures—spirals, mandalas, and other forms that mirror the natural order.

In the language of string theory, each vibrating string creates its own geometry, its own unique pattern of existence. This sacred geometry is the blueprint of the universe, the invisible architecture that underlies all form and function. To perceive this geometry is to see the cosmos as a work of art, a masterpiece of resonance where every vibration contributes to the whole.

To the mystic, this geometry is a language through which

the divine speaks, a bridge between the material and the spiritual. Each vibration, each sound, is a symbol of the unity that pervades all things, a reminder that life is not chaos but composition, a song whose beauty lies in its harmony.

Life as a Dance of Frequencies

If the universe is a symphony, then life itself is a dance of frequencies, a melody composed of countless notes, each unique yet interconnected. Each of us vibrates at our own frequency, creating a personal song that weaves into the greater composition of existence. This individuality does not separate us; it unites us, as each melody contributes to the symphony of the whole.

The mystic understands life as a journey of resonance, a path where each thought, each action, each experience is a note in the song of the soul. To live in harmony with this cosmic dance is to align oneself with the rhythm of the universe, to feel the pulse of creation, to move with the flow of existence. This alignment brings a sense of peace, a joy that arises from knowing one's place in the symphony of life.

The Silence That Holds the Music

In every melody, there is a silence between notes, a stillness that gives the music its shape, its meaning, its soul. The universe, too, is shaped by this silence, a space that holds the vibrations of existence, a void that gives birth to sound. In string theory, this silence is the vacuum in which the strings vibrate, the emptiness that contains the fullness of creation.

The mystic experiences this silence as the presence of the infinite, a stillness that pervades all things, a peace that is both emptiness and fullness. To listen to the silence between notes is to hear the voice of the cosmos, to touch the essence of being, to feel the presence of the divine. This silence is not absence but potential, a space where the music of the spheres is born.

Unity in Diversity

In the cosmic symphony, each note is distinct, each vibration unique, yet all are part of the same song. This unity in diversity is the essence of the universe, a harmony that embraces difference, a oneness that celebrates multiplicity. String theory reveals that each particle, each force, each form is a variation on a single theme, a note in the music of the spheres.

The mystic sees this unity as the essence of love, a force that binds all things in harmony, a presence that flows through each being, each moment, each breath. To live in this harmony is to honour the beauty of difference, to see each soul as a unique expression of the same light, a note in the cosmic song. In this vision, life becomes a celebration of unity, a dance of diversity, a symphony where every voice is cherished, every note essential.

The Heartbeat of the Universe

In the depths of the cosmos, in the silence of the soul, there is a heartbeat—a rhythm that connects the finite with the infinite, the temporal with the eternal. This heartbeat is the essence of the cosmic symphony, a reminder that we are not isolated fragments but integral parts of a greater whole.

To feel this heartbeat is to sense the pulse of creation, to recognize that life is not a solitary journey but a shared experience, a symphony where each of us is both musician and audience. In this heartbeat, we find our place in the universe, a place that is both humble and profound, both ordinary and sacred.

The Infinite Song

The cosmic symphony is not a finished composition but an ongoing melody, a song that evolves with each moment, each vibration, each breath. In every instant, the universe creates itself anew, each string vibrating with the energy of creation, each note adding to the infinite harmony.

To live in tune with this melody is to embrace the flow of existence, to see life as a part of the music of the spheres. This is the journey of the mystic, the path of the soul—a journey without end, a melody without finality, a song that celebrates the boundless beauty of existence.

In the infinite symphony of the cosmos, we are not merely listeners but participants, not merely observers but creators. To embrace this truth is to live in harmony with the universe, to hear the music in the silence, to feel the resonance in the stillness, to dance to the rhythm of the infinite. This is the gift of the cosmic symphony—a life lived in tune with the divine.

CHAPTER 10: THE ALCHEMIST'S DREAM – TRANSMUTATION IN THE QUANTUM FIELD

In the hidden realms of nature, where atoms and elements spin in silent symphony, there exists a possibility as old as human wonder: the power of transmutation, the transformation of one state into another, of base into sublime. In ancient times, alchemists sought this power, hoping to turn lead into gold, to refine the common into the extraordinary. Yet, beyond the literal quest for physical gold lay a deeper purpose, a journey of the soul toward inner transformation, a metaphor for the change from ignorance to wisdom, from limitation to liberation.

Quantum physics opens a new doorway into this dream, revealing a universe where particles can shift states, where matter and energy exchange identities, where the boundaries between elements dissolve. In this quantum world, change is not only possible but intrinsic, a reminder that transformation is woven into the fabric of existence, a natural unfolding of potential. Here, we explore the alchemist's dream not as an ancient curiosity but as a profound truth about the universe—and about ourselves.

The Quest for Gold: Alchemy as Inner and Outer Transformation

Alchemy, as the ancient art of transformation, was more than a pursuit of material wealth; it was a journey of the soul, a path of purification, an attempt to refine the self along with the substance. To transmute lead into gold was, in essence, to elevate the self from base desires to noble aspirations, to reveal the divine within the human, the eternal within the temporal. The mystics of old knew that the philosopher's stone, the mythical agent of transformation, was a symbol of the soul's potential, a metaphor for the power to change, to grow, to become.

In the quantum realm, we see a parallel to this quest, a reminder that transformation is not confined to the physical but is a process that permeates all levels of being. Particles shift states, atoms merge and separate, energy becomes matter, and matter returns to energy in an endless cycle of creation and dissolution. To understand this quantum transmutation is to see that the universe is a dynamic field of possibility, a space where everything is in a state of becoming, where each moment is an opportunity to transform, to transcend, to touch the infinite.

The Quantum Field: A Playground of Possibilities

The quantum field is not a static stage but a playground of possibilities, a space where particles exist in multiple states, where identity is fluid, where boundaries blur. In this field, particles are not fixed entities but potentials, shifting forms that respond to observation, to interaction, to the dance of energy that flows through them. This fluidity reveals a truth that resonates with the alchemist's dream: that reality itself is mutable, that transformation is not only possible but inherent, a process that lies at the heart of existence.

For the alchemist, the laboratory was a sacred space, a place where spirit and matter met, where intention could shape reality, where the mind's focus could alter the substance of the world. The quantum field, too, is a space of sacred transformation, a field where observation shapes reality, where consciousness plays a role in the unfolding of existence, where each particle holds within it the possibility of change. To work with this field is to embrace the power of transmutation, to recognize that we are co-creators in the dance of life, that our thoughts, intentions, and actions are part of the alchemy of the cosmos.

The Philosopher's Stone: The Power to Transform

In alchemy, the philosopher's stone was the key to transmutation, the agent of transformation that could turn lead into gold, mortal into divine, finite into infinite. This stone was not a literal object but a symbol, a representation of the soul's potential, a reminder that we each hold within us the power to transform, to change, to transcend the limitations of the self. The philosopher's stone is the mind awakened, the heart opened, the spirit attuned to the resonance of the divine.

In quantum mechanics, we find a similar power in the observer effect, the capacity of consciousness to influence the state of reality, to collapse the wave of potential into a specific outcome. The philosopher's stone, then, is the ability to choose, to direct the flow of possibility, to shape the self and the world through the power of awareness. This is the alchemy of the soul, a process that invites us to become the philosopher's stone itself, to embody the power of transformation, to be the agents of our own becoming.

Transmutation as a Journey of the Soul

Just as the alchemist sought to refine base metals into gold, so too does the soul seek to refine itself, to transform its fears into love, its doubts into faith, its ignorance into wisdom. This inner transmutation is the true work of the alchemist, the process of turning the "lead" of the ego into the "gold" of the spirit, a journey that leads not to material wealth but to the treasure of self-realization, to the discovery of one's true nature as an expression of the divine.

The quantum field mirrors this inner journey, showing us that transformation is a natural process, a flow that occurs not by force but by alignment, by attunement to the deeper frequencies of existence. To align with this flow is to surrender to the journey of the soul, to trust that each experience, each challenge, each joy, is a step in the process of becoming, a step that brings us closer to the realization of our own divine potential. In this journey, there is no failure, no loss, only the unfolding of the self, the continual refinement that leads to the alchemist's gold.

The Dance of Energy and Matter: The Universe as Alchemist

In the quantum world, matter and energy are not separate but interchangeable, different expressions of the same underlying reality. Particles transform into waves, waves into particles, energy becomes mass, and mass returns to energy in an endless dance of transformation. This dance is the alchemy of the universe, a reminder that nothing is fixed, that all things are in a state of flux, that change is the essence of being.

The mystic sees this dance as the breath of the divine, a rhythm that flows through all things, a movement that transcends form and reveals the unity of existence. To live in harmony with this dance is to embrace the power of transformation, to see that life is a process of becoming, a journey where each moment holds the potential for change, for growth, for transcendence. This is the path of the alchemist, the way of the soul—a journey that leads not to the attainment of a final state but to the continual expansion of awareness, the ever-deepening realization of one's own divine nature.

The Art of Self-Transformation: The Alchemist Within

To be an alchemist is to be an artist of the soul, a creator who shapes reality not through force but through awareness, through the power of intention, through the alignment of the self with the greater harmony of the universe. The alchemist understands that true transformation begins within, that to change the world, one must first change oneself, that the outer and the inner are reflections of each other, that the lead and the gold are two aspects of the same essence.

The mystic sees this art of self-transformation as the highest form of alchemy, a process that refines the soul, that purifies the mind, that opens the heart to the beauty of existence. This is the journey of becoming, the path of self-realization, a journey that leads to the discovery of one's true nature, a nature that is both human and divine, both finite and infinite, both temporal and eternal. To walk this path is to become the alchemist, to embody the philosopher's stone, to live as a being of transformation, a being who creates not through will but through love.

The Fire of Change: Embracing the Crucible

In alchemy, the crucible is the vessel in which transformation occurs, the container that holds the substance as it is refined by fire, as it releases its impurities, as it reveals its essence. In the journey of the soul, the crucible is life itself, the experiences that shape us, the challenges that test us, the moments that call us to change. To embrace the crucible is to embrace the fire of transformation, to accept that true growth requires sacrifice, that the path to wisdom is paved with the willingness to let go of what no longer serves, to surrender the "lead" of the ego in order to discover the "gold" of the spirit.

In the quantum field, this crucible is the space of potential, the field where particles are transformed, where matter and energy meet, where existence itself undergoes continual renewal. To live in this field is to live in the presence of the infinite, to recognize that each moment holds the power to transform, to release, to transcend. This is the alchemist's dream, the vision of a life lived in harmony with the process of becoming, a life that celebrates change as the essence of existence, a life that honours the beauty of transformation.

Becoming the Philosopher's Stone: The Gift of Transmutation

The alchemist's dream is not merely to change the world but to become the change, to embody the process of transformation, to live as the philosopher's stone, a being who holds within the power to transmute, to elevate, to create. In this vision, each of us is a potential philosopher's stone, a soul capable of transformation, a consciousness capable of choosing, a being capable of becoming.

To become the philosopher's stone is to awaken to one's own power, to see that transformation is not an external process but an inner journey, a journey that leads to the realization of one's own divine nature. This is the essence of alchemy, the recognition that the true gold is not material but spiritual, that the true transmutation is not of lead into gold but of self into soul, of fear into love, of limitation into liberation.

The Endless Journey of Transformation

And so, we return to the alchemist's dream, to the recognition that life itself is a process of transmutation, a journey where each moment is an opportunity to change, to grow, to become. In the quantum field, we see a reflection of this journey, a reminder that transformation is not a destination but a way of being, a state of openness, a willingness to embrace the unknown, a courage to step into the fire of change.

To live as an alchemist is to live with an awareness of possibility, to see each experience as a chance to transform, each challenge as a step in the journey of the soul, each moment as a doorway to the infinite. This is the gift of transmutation, the beauty of a life lived in harmony with the flow of existence, the joy of a journey that has no end, only the endless unfolding of the self in the light of awareness.

Embracing the Alchemist Within

In the end, to embrace the alchemist within is to embrace the truth of who we are—beings of transformation, beings of potential, beings who are both human and divine, both finite and infinite. This is the wisdom of the mystic, the vision of the philosopher, the insight of the physicist—a recognition that life is not fixed but fluid, a journey that invites us to change, to grow, to become.

To live as an alchemist is to live as a creator, a lover, a dreamer, a soul who sees the world not as a prison but as a playground, a space of possibility, a field of becoming. This is the path of transmutation, the way of the soul, a journey that leads not to an end but to an endless beginning, a journey that is, in truth, the alchemist's dream.

CHAPTER 11: THE UNSEEN DIMENSIONS – BEYOND THE PHYSICAL REALM

Beneath the visible, tangible world lies a vast expanse of mysteries, a multidimensional reality concealed behind the thin veil of perception. These unseen dimensions, theorized by modern physics and intuited by mystics across millennia, point to a universe far more expansive and intricate than the one we encounter with our senses. Where science speaks of hidden planes, compactified dimensions, and parallel realities, mysticism ventures into realms of spirit, energy, and consciousness—spaces where the limits of physicality dissolve, and the infinite reveals itself.

To peer beyond the veil is not merely an intellectual exercise but a profound act of courage and wonder. It is to step outside the boundaries of what is known, to explore the layered nature of existence, and to embrace the truth that what we see is but a fraction of the cosmos. These hidden dimensions invite us to consider that reality is not solid but fluid, not finite but boundless, a dance of the visible and the invisible, the material and the metaphysical.

The Possibility of Higher Dimensions

Modern physics suggests that the dimensions we inhabit—three of space and one of time—are but a subset of a far richer framework. String theory, a daring venture into the nature of reality, posits the existence of up to eleven dimensions, many of which are hidden, folded into spaces so infinitesimal they evade detection. These unseen dimensions may hold the answers to the universe's deepest mysteries, from the nature of gravity to the unification of forces that govern the cosmos.

For the mystic, such hidden dimensions resonate deeply, though their language is different. These spaces are not measured in numbers but felt in the heart, not mapped but experienced. They are realms where time bends and form dissolves, where the soul encounters energies and presences that exist beyond the material world. To explore these dimensions is to understand that reality is not confined to what is observable but is a layered expanse, each layer revealing deeper truths about existence, connection, and the divine.

The Veil of Perception: What Lies Beyond the Senses

Our senses, though wondrous, are narrow in scope. They filter reality, presenting us with a version that is manageable but incomplete. What we perceive as solid and finite is, in truth, a fragment of an infinitely more complex universe. The hidden dimensions, both scientific and mystical, remind us of how much lies beyond our sensory grasp.

The mystic approaches this veil not with scepticism but with reverence, understanding it as a sacred boundary that invites exploration. Through meditation, stillness, and contemplation, one can transcend the limitations of sight and touch, entering realms where the unseen becomes felt, where energy, light, and vibration take precedence over matter. In these moments, the veil lifts, revealing a reality teeming with presence, depth, and interconnectedness—a world not bound by physicality but illuminated by the infinite.

The Multiverse: Infinite Realities Beyond Our Own

Theoretical physics has expanded our understanding of the universe to include the possibility of a multiverse: an infinite collection of parallel universes, each a unique expression of reality. In this view, our universe is not the only melody in the cosmic symphony but one of countless variations, each universe governed by its own laws, its own dimensions, its own unfolding story.

Mystics, too, have long spoken of multiple realms of existence. These are not physical worlds but planes of consciousness—levels of reality where the soul journeys, where spirit and energy manifest in infinite forms. To glimpse these realms is to see that life does not end at the edge of the visible but stretches endlessly, encompassing dimensions where the self-encounters the divine, where truth takes on countless expressions, and where the infinite reveals itself in infinite ways.

The Inner Dimensions: The Universe Within

As vast as the external cosmos may be, the mystic reminds us that an equally infinite expanse exists within. The self is not merely a collection of thoughts and memories but a microcosm of the universe, a being composed of layers upon layers of depth, each one mirroring the multidimensional nature of reality.

Within the inner landscape, there are dimensions of thought, emotion, intuition, and awareness—spaces where the soul explores its own vastness. Meditation becomes the vehicle for this journey inward, a practice that reveals fields of light, expanses of silence, and energies that transcend the physical. To journey within is to recognize that the boundaries between the internal and the external dissolve, that the universe is not only out there but also in here, that we are beings whose essence spans dimensions both seen and unseen.

The Quantum Realm: Dimensions of Possibility

Quantum mechanics reveals a realm of reality that defies the fixed and static. In the quantum field, particles exist in states of probability, inhabiting multiple possibilities simultaneously until observed. This fluidity of existence points to dimensions that are not physical in the traditional sense but are spaces of potential, realms where reality is shaped not by certainty but by choice and consciousness.

To the mystic, this quantum fluidity echoes the spiritual truth that existence is not predetermined but creative, that each moment holds infinite possibilities, shaped by the intentions of the soul. This is the realm of co-creation, where thought becomes form, where desire shapes reality, where consciousness is the architect of existence. To explore this dimension is to understand that we are not passive observers but active participants in the unfolding of the universe, creators in the dance of becoming.

The Light and the Shadow: The Duality of the Unseen

In the hidden dimensions of reality, light and shadow intertwine, creating a dynamic interplay that shapes existence. These dimensions are not separate from us but are interwoven with our lives, influencing our thoughts, emotions, and actions in ways that are often subtle, sometimes profound.

The mystic senses the presence of these unseen forces in moments of intuition, in the quiet whispers of insight, in the inexplicable connections that thread through life. To live with an awareness of these dimensions is to embrace both the light and the shadow, to see that existence is not a duality but a dance—a movement between the visible and the invisible, the known and the unknown, the material and the spiritual.

The Veil as Invitation: Stepping into the Mystery

The unseen dimensions are not obstacles but invitations, beckoning us to expand our understanding of reality.

The veil between the known and the unknown is not a wall but a threshold, a liminal space that calls us to step beyond the familiar and encounter the infinite.

For the mystic, this veil is sacred—a boundary that both separates and connects, that hides and reveals. To approach the veil is to feel the presence of something greater, to sense the vastness that lies beyond sight, to understand that the universe is a doorway into the transcendent. This invitation is not about escaping the world but deepening our experience of it, seeing each moment as a reflection of the unseen, each breath as a bridge to the infinite.

The Journey Beyond: The Path of Expansion

To explore the unseen dimensions is to embark on a journey of expansion, a journey that deepens our understanding of self and reality. This path is not about leaving the physical behind but integrating it with the spiritual, recognizing that the visible and the invisible are two sides of the same coin.

The mystic sees this journey as one of awakening, a process of peeling back the layers of illusion to reveal the truth beneath. It is a path that requires courage, openness, and a willingness to embrace mystery—a path that leads not to answers but to deeper questions, not to conclusions but to communion with the infinite.

Living with the Unseen

To live with an awareness of the unseen dimensions is to live in harmony with the mystery of existence. It is to recognize that life is not a series of random events but a sacred unfolding, a dance that weaves together the seen and the unseen, the physical and the metaphysical.

This awareness invites us to honour the depth of each moment, to see the infinite within the finite, to live as beings who are part of both the material and the spiritual worlds. To embrace the unseen is to embrace the fullness of life, to walk through the world with a sense of wonder, and to journey ever deeper into the heart of existence.

In the unseen dimensions, we find not only the mysteries of the cosmos but the truth of who we are—beings of light, beings of depth, beings whose essence transcends the boundaries of space and time. This is the gift of the unseen: a life lived in communion with the infinite, a journey into the boundless, a recognition that the universe is far more than it seems.

CHAPTER 12: THE DIVINE UNCERTAINTY – HEISENBERG AND THE LIMITS OF KNOWING

At the core of existence, nestled within the unseen fabric of reality, lies uncertainty—a dance so delicate and profound that it disrupts our instinct for precision, our yearning for control, our insatiable hunger to "know." Werner Heisenberg, physicist and philosopher, illuminated this truth through his groundbreaking uncertainty principle, a revelation that shattered the foundations of classical certainty. According to his principle, the more precisely one measures a particle's position, the less certain its momentum becomes, and vice versa. This was not a flaw in human instrumentation but a fundamental truth of nature—a universe that conceals as much as it reveals, guarding its mysteries with the elegance of uncertainty.

For the mystic, Heisenberg's insight is not merely a scientific observation but a divine metaphor, an echo of ancient wisdom. It speaks to the ineffable nature of the cosmos, where truths elude capture and mysteries beckon not to be solved but to be lived. The principle resonates as a reminder that life's greatest depths cannot be dissected by logic alone, that some realities are beyond the intellect's grasp, and that true understanding lies in surrendering to the wonder of what cannot be known.

The Edge of Knowledge: Embracing the Unknown

Heisenberg's uncertainty principle reveals a poignant boundary—an edge where certainty dissolves into ambiguity, and the ordered scaffolding of thought gives way to the boundless unknown. This edge is not a flaw in our understanding but an essential feature of reality. It reminds us that existence is not a fixed structure to be mapped but a shifting interplay of probabilities, a symphony of interactions that respond to the act of observation itself.

For the mystic, this edge of knowledge is a sacred threshold. It is the place where the ego falters, where the mind pauses, and where the soul begins to listen. At this boundary, one senses the presence of something vast and eternal, a whisper of the infinite that cannot be pinned down or defined. Here, wisdom arises not from mastery but from humility, from the acknowledgment that there are truths we are meant to experience, not to control. To stand on this precipice is to feel the paradoxical freedom of letting go, to encounter a vastness that can only be embraced by the heart.

The Paradox of Knowing and Unknowing

The uncertainty principle encapsulates a paradox: the act of knowing one aspect of reality inevitably obscures another. This interplay reflects a deeper truth—that knowledge is inherently incomplete, that to define is also to limit, that what we grasp in the hand of reason slips through the fingers of intuition. For every particle measured, there is a wave unmeasured; for every certainty claimed, there is a mystery relinquished.

Mystics have long understood this paradox as a central feature of spiritual wisdom. They teach that true knowing arises not from accumulation but from release, not from answers but from openness to questions. To seek absolute certainty is to confine reality to a box too small for its infinite nature. Instead, the mystic invites us to dwell in the space of unknowing, to embrace the paradox of a truth that cannot be fully spoken, a mystery that reveals itself only when we stop trying to grasp it.

This unknowing is not ignorance; it is a higher form of understanding, a way of encountering the divine that transcends the limitations of thought. It is the quiet realization that the deepest truths are not solved like puzzles but lived like poetry, that they unfold not in conclusions but in wonder.

The Mystery as Teacher: Lessons in Humility and Awe

Heisenberg's principle serves as a humbling teacher, reminding us that reality resists domination, that existence is not a riddle to be conquered but a mystery to be revered. For the mystic, this humility is the beginning of wisdom, a sacred invitation to approach life with awe rather than control. To accept uncertainty is to bow before the vastness of existence, to admit the limits of our knowledge, and to open ourselves to the beauty of what lies beyond.

This humility births a sense of wonder, a childlike reverence for the miracle of being. In the face of uncertainty, the mystic sees not a void but a fullness—a realm rich with possibilities, a cosmos alive with meaning that cannot be reduced to equations. Mystery becomes the greatest teacher, urging us to trust in the unseen, to honour what we cannot understand, and to find joy in the infinite complexity of the world.

The Dance of Control and Surrender

At the heart of Heisenberg's insight is a revelation about control: the more we attempt to pin reality down, the more it slips through our grasp. This truth challenges our desire to master the world, to fix it in place, to extract certainty from its ever-changing essence. Control, it seems, is an illusion—a fleeting mirage in a universe that thrives on flux and fluidity.

The mystic sees this dance of control and surrender as the essence of life. To cling tightly to certainty is to resist the natural flow of existence, to fight against the current of the infinite. True freedom lies in surrendering to the dance, in letting go of the need to control, in trusting that life unfolds with a wisdom far greater than our own. This surrender is not passivity; it is an active opening to the mystery, a willingness to move with the rhythm of the unknown, to find harmony in the flow of what is.

The Divine as Mystery: Presence Beyond Understanding

For the mystic, the divine is not a puzzle to be solved but a presence to be felt—a mystery that pervades all things, defying the boundaries of thought and language. This divine presence is not something to be grasped but something to be experienced, a truth that can only be known in the silence of the soul.

In moments of stillness, the mystic senses the divine as an ineffable presence, a love that transcends understanding, a light that shines even in the shadows of unknowing. This presence is not diminished by uncertainty; it is revealed through it. To encounter the divine as mystery is to let go of the need for definitions, to see that the sacred cannot be confined to the limits of reason, to recognize that the infinite is not something to be captured but something to be lived.

Living with Divine Uncertainty: The Freedom of Openness

To live with divine uncertainty is to live with a profound freedom—a freedom born not of answers but of openness, not of control but of trust. In the space of not-knowing, there is a lightness, a sense of possibility, a recognition that life is not a series of fixed outcomes but an unfolding process, a journey into the infinite.

The mystic embraces this freedom as a path of liberation, a release from the burden of needing to know, a letting go that allows the self to expand. In this openness, life becomes a dance of discovery, a continuous unfolding where each moment is a revelation, each experience a doorway to the sacred. This is the freedom of divine uncertainty—the freedom to be present, to explore, to wonder, to live fully in the light of the unknown.

The Gift of Divine Uncertainty: Embracing the Mystery

Heisenberg's uncertainty principle is more than a scientific truth; it is a profound gift, a reminder that life's beauty lies not in its certainties but in its mysteries. To embrace this gift is to release the need for control, to find peace in the unknown, to see that the limits of knowledge are not barriers but invitations.

In this embrace, we discover a deeper truth—that life is not a problem to be solved but a miracle to be experienced, a sacred unfolding that leads us ever deeper into the heart of existence. This is the gift of divine uncertainty—a gift that frees the soul, expands the heart, and illuminates the mind.

The Infinite Embrace of Mystery

In the depths of uncertainty, we find the essence of the divine—a presence that transcends understanding, a love that defies explanation, a mystery that holds all things. To live in this embrace is to walk a path of wonder, to see that life is not a destination but a journey, not a certainty but an unfolding.

This is the heart of divine uncertainty: a call to trust in the unknown, to surrender to the infinite, to live with a sense of wonder that opens the soul to the boundless beauty of existence. It is a path that leads not to answers but to communion, not to mastery but to harmony, a path that reveals the truth of who we are—beings of mystery, beings of light, beings forever expanding into the infinite unknown.

CHAPTER 13: WAVES OF CONSCIOUSNESS – MIND AS QUANTUM ENERGY

Within the silent chambers of awareness, an eternal resonance whispers—a vibration both intimate and infinite, weaving through the mind's recesses and the cosmic expanse. This resonance, delicate yet profound, suggests that consciousness is not confined to the biological corridors of the brain but is instead a wave, a field, an energy that pulses through the quantum fabric of existence.

If matter is energy bound into form, might not thought, too, be energy unbound, a ripple across dimensions, a force that shapes and is shaped in return? Physics unveils a universe not of discrete particles but of undulating waves, interacting in a symphony of probabilities. In this paradigm, consciousness reveals itself not as a static entity but as an ever-flowing current—an energetic dance entwined with the quantum field, echoing through the architecture of the cosmos.

To view consciousness as a wave is to step beyond the narrow corridors of mechanistic thought, to perceive mind as a force—dynamic, expansive, and indivisible from the rhythms of creation itself. It invites us to see the self not as a solitary island of awareness but as a node within a universal network, a vibrant melody within an infinite symphony.

The Quantum Field of Thought: Energy in Motion

At the quantum level, the universe is a sea of probabilities—a realm where particles flicker into and out of existence, their trajectories shaped by the invisible waves of potential. In this reality, every particle exists not in isolation but as part of a vast, interconnected field. Similarly, consciousness can be imagined as a field of awareness, a space where thoughts arise as waves, flowing outward to influence the seen and unseen.

Each thought, like a quantum fluctuation, sends ripples across the fabric of mind, resonating with energy that transcends its origin. The mystic understands this instinctively, feeling how thoughts carry vibrations that extend beyond the self, touching others, altering environments, and shaping possibilities. To embrace this vision is to recognize that thought is not merely internal chatter but a force of creation, a pulse within the quantum sea. Each intention becomes a wave, each moment of awareness a thread in the great fabric of existence.

The Unified Field of Consciousness: Minds as One

If waves of thought extend outward, then the boundaries between minds begin to blur, suggesting a deeper unity —a field of consciousness that binds all beings. In the quantum world, entangled particles remain connected across vast distances, responding to one another instantaneously, as if distance were an illusion. Might our minds, too, be entangled, resonating within a single, unified field?

The mystic feels this interconnectedness in moments of profound empathy, in the shared silence of meditation, in the unspoken connection that transcends words. These moments reveal that we are not isolated selves but expressions of a greater awareness, strands in a collective web that vibrates with the thoughts, feelings, and intentions of all beings. To live with this awareness is to honour the shared mind, to understand that every thought contributes to the whole, that we are part of a vast, interwoven consciousness that pulses with life's infinite potential.

Resonance of Intention: Thoughts as Waves of Power

If thoughts are waves, they carry frequencies, vibrations that shape the contours of reality. Just as waves in the quantum field interact—amplifying, cancelling, creating patterns of infinite complexity—so too do our thoughts resonate within the field of consciousness, weaving patterns that define our experiences and influence the world around us.

The mystic knows that intention is a powerful vibration, a force that aligns thought with creation. In stillness, the mind becomes a calm lake, where each thought is a ripple that touches the depths and extends outward, carrying energy that transforms. This resonance is sacred, a reminder that each thought is a seed, a frequency that shapes not only the self but the collective. To live with this understanding is to wield thought with care, to recognize its creative and transformative power, to align intention with love, purpose, and the infinite.

The Mind's Dance: Fluidity as Essence

In the quantum realm, particles and waves are not static; they shift, they dance, they spiral in response to unseen forces. This is not chaos but a higher order, a harmony that reflects the flow of existence itself. Consciousness, too, is a dance—a fluid movement of awareness that responds to the rhythms of life, to the currents of thought, to the invisible pull of intention and attention.

The mystic sees this dance of consciousness as the essence of being, a rhythm that calls us to let go of rigidity, to flow with life's ever-changing melody. To live in alignment with this dance is to embrace the freedom of the mind, to move with the currents of creativity, to see the self as dynamic, evolving, and expansive. In this dance, the mind becomes a wave, a quantum energy that pulses with the rhythm of the cosmos, reflecting the infinite beauty of existence.

Collective Consciousness: The Shared Symphony of

Thought

If consciousness is a field, then we are not separate islands but notes within a collective symphony—a vast mind that holds the thoughts, dreams, and memories of all beings. This collective consciousness is not an abstraction but a palpable presence, a resonance that flows through cultures, histories, and civilizations, shaping the shared human experience.

The mystic feels this collective mind in moments of unity, in the shared heartbeat of humanity, in the sense that we are all part of a single story, a single soul. This awareness brings with it a profound responsibility: to contribute to the field with thoughts of love, wisdom, and compassion, to recognize that each intention ripples outward, touching the whole. To live with this awareness is to honour the interconnectedness of all life, to see oneself as both an individual and a part of the infinite.

The Mind as Creator: Consciousness Shapes Reality

In the quantum field, observation alters reality; the act of seeing shapes what is seen. Consciousness, then, is not passive but active—a force that co-creates the universe through perception, intention, and focus. The mystic understands this creative power as a divine gift, a recognition that the mind is not merely a mirror but a flame, a spark that ignites possibilities and shapes the world.

To live as a creator is to embrace the responsibility of consciousness, to see thought as a tool of transformation, to align the mind with the highest good. This is not an act of control but of co-creation, a partnership with the universe that honours the sacred dance of being. In this vision, the mind is a wave that flows through the quantum field, a force that resonates with the infinite, a creator in the unfolding story of existence.

The Soul's Resonance: Consciousness as Divine Vibration

At the deepest level of consciousness, there is a vibration—a song that resonates with the divine, a frequency that connects the self to the infinite. This is the soul's resonance, the essence of awareness that flows through all beings, a melody that unites the individual with the universal, the finite with the eternal.

The mystic hears this song in the silence of meditation, in the beauty of nature, in the quiet knowing that arises when the mind is still. This resonance is not separate from us; it is within us, a vibration that reflects the divine presence, a wave that flows through the heart of existence. To live with an awareness of this resonance is to feel the divine within, to recognize that consciousness is not apart from creation but is itself a vibration of the sacred.

The Infinite Ocean of Awareness

Consciousness, like the quantum field, is an ocean—vast, boundless, and interconnected. Each thought, each feeling, each moment of awareness is a wave within this ocean, a ripple that flows outward, connecting the self to the whole. To see consciousness in this way is to recognize the infinite nature of the mind, to understand that we are not confined by the physical but are expansive, resonant, and infinite.

This is the gift of consciousness as quantum energy—a reminder that we are beings of light, beings of vibration, beings who resonate with the essence of the cosmos. To embrace this vision is to live with an awareness of the infinite, to see each moment as a wave in the ocean of consciousness, to honour the interconnectedness of all life.

Embracing the Waves of Mind

To embrace the waves of consciousness is to embrace the truth of existence—that we are both individuals and part of the whole, both finite and infinite, both human and divine. This is the wisdom of the mystic, the insight of the seeker, the revelation of the quantum field—a recognition that life is a field of resonance, a dance of waves, a song of the soul.

To live in this awareness is to honour the waves of thought, to see consciousness as a sacred force, to recognize that each moment is an opportunity to create, to connect, to resonate with the infinite. This is the beauty of mind as quantum energy, the joy of a life lived in harmony with the waves of existence, the gift of a journey that leads ever deeper into the heart of being.

CHAPTER 14: THE SECRET CODE – SACRED GEOMETRY AND QUANTUM PATTERNS

Hidden within the invisible lattice of existence lies an exquisite code, a cryptic hymn composed not of words, but of shapes, harmonics, and rhythms that breathe form into the formless. It is the silent architect of all we perceive: the symmetry of a snowflake, the spiral of a nautilus shell, the unfurling of fern fronds, and the swirling arms of distant galaxies. Known as sacred geometry, this ancient and universal language reveals the harmonious undercurrent of creation, a blueprint whispered by the divine into the very fabric of existence.

Modern physics, too, uncovers a startling resonance with this ancient wisdom. Beneath the randomness of subatomic chaos, quantum mechanics unveils symmetries and patterns that echo sacred proportions. The quantum field, with its waves and lattices, mirrors the geometric perfection of nature's designs, suggesting that the universe itself is a symphony of structure, a melody sung by equations and ratios older than time.

This chapter seeks not merely to observe these patterns but to dwell in their poetry, to expand the lens of understanding until the infinite relationships between sacred geometry and quantum mechanics reveal the cosmos as a singular, radiant harmony—a code of profound beauty that connects the visible and the unseen, the material and the transcendent.

Geometry as the Script of Creation: Universal Archetypes

To perceive the world through the lens of sacred geometry is to recognize that shapes are not mere aesthetic curiosities but archetypes—eternal symbols that underlie all that is seen and unseen. These shapes, whether the perfect circle, the balanced triangle, or the intricate interlocking of the flower of life, are not merely designs but the DNA of existence itself, blueprints that govern the formation of everything from atoms to galaxies.

The circle, infinite in its continuity, symbolizes unity and wholeness. The triangle, stable and unwavering, embodies balance and strength. The spiral, found in DNA and hurricanes alike, speaks of growth, evolution, and the infinite unfolding of life's mysteries. The hexagon, ubiquitous in the honeycomb and snowflake, manifests efficiency and interconnection, its angles expressing the cosmic dance between individuality and unity.

When we look to the patterns of life—bees constructing

hexagonal combs, flowers forming petals in Fibonacci sequences, rivers carving fractal paths through the earth—we see that sacred geometry is not merely symbolic. It is the invisible scaffolding of nature, the silent architect orchestrating life's designs with a precision that transcends randomness.

Quantum Harmony: The Hidden Order of Reality

Modern physics reveals an unexpected kinship with this ancient wisdom. At the quantum level, reality is not the chaotic swirl of randomness it appears to be. Instead, particles move according to elegant symmetries, wave functions curve in harmonious oscillations, and the quantum lattice forms geometric fields of energy.

For instance, electrons orbit nuclei in discrete patterns, their orbits governed by probabilities that echo the proportions of sacred geometry. Subatomic particles, too, interact with one another according to principles of symmetry and resonance, creating patterns as intricate as the petals of a rose or the tessellation of a honeycomb.

The mystic and the physicist, often walking divergent paths, converge in their recognition of a deeper order. The quantum world, with its lattice of probabilities, suggests that even at its most fundamental level, reality is structured, interconnected, and imbued with meaning. Sacred geometry, once dismissed as mystical abstraction, finds its reflection in the quantum realm, where the universe speaks the same language in smaller, subtler tones.

The Flower of Life: A Gateway to Infinity

Among the symbols of sacred geometry, the Flower of Life holds a place of profound reverence. Composed of interlocking circles, this pattern forms a radiant web of interconnectedness, a map of unity that echoes across cultures and ages. It has been discovered etched into the stones of ancient temples and whispered through the sacred art of countless civilizations, from the Egyptian priests to the scholars of Renaissance Europe.

In the quantum realm, the Flower of Life finds its reflection in the interconnected nature of particles, the entanglement that binds them across time and space. Just as the circles of the Flower of Life overlap to form a harmonious whole, so too do the quantum fields overlap, their energies merging, resonating, creating a seamless web that holds the universe together.

To contemplate the Flower of Life is to glimpse infinity. Each circle is both complete in itself and part of the greater whole, a microcosm of creation that mirrors the macrocosm. It teaches us that separation is an illusion, that all things are one, that the cosmos itself blooms like a flower, forever unfurling in infinite patterns of possibility.

The Golden Ratio: The Pulse of Harmony

The Golden Ratio, or phi, is an irrational number woven into the very fabric of existence. Found in the spirals of galaxies, the proportions of the human body, the branching of trees, and even the beating of the heart, it is a ratio that speaks of harmony, balance, and beauty. Phi, approximately 1.618, is not merely a number; it is a principle that governs the growth of life and the unfolding of the universe.

In the quantum field, the Golden Ratio resonates in the harmonics of waves, the proportions of particle interactions, the symmetry of energy distributions. Just as ancient architects and artists used phi to create works of enduring beauty, so too does the cosmos weave this ratio into its designs, a silent signature of the divine.

The mystic sees the Golden Ratio as a bridge between the finite and the infinite, a reminder that beauty is not an accident but a reflection of deeper truths. To live in harmony with this ratio is to align oneself with the rhythms of existence, to feel the balance that flows through all things, to honour the unity that underlies diversity.

Fractals: The Infinite Repeated in the Finite

The fractal is a geometric shape that repeats itself at every scale, revealing infinite complexity within simplicity. Found in the spiralling shells of snails, the branching of rivers, and the shapes of clouds, fractals are a testament to the recursive nature of creation. Each part contains the essence of the whole, and the whole is mirrored in its smallest fragments.

Fractals echo through the quantum world as well. Fields of energy ripple in self-similar patterns, particle interactions create recursive waves, and the universe itself expands in fractal-like dimensions. These patterns suggest that existence is not linear but cyclical, not finite but infinite, a design that reflects the unity of the macrocosm and the microcosm.

To contemplate fractals is to witness the infinite within the finite, to see that each moment, each breath, each thought reflects the whole. It is to understand that life itself is fractal, a journey of continuous creation, a pattern that repeats and evolves, a dance that reflects the divine.

Sacred Geometry as the Divine Blueprint

Sacred geometry reveals the cosmos as a work of art, a masterpiece shaped by a divine intelligence that infuses every atom, every star, every soul. It suggests that the universe is not a chaotic accident but a purposeful creation, a reflection of principles that transcend time and space.

In the quantum realm, we find the same evidence of intention—a universe built on ratios, symmetries, and patterns that resonate with the harmonies of sacred geometry. The physicist and the mystic, though often speaking different languages, are both drawn to this truth: that reality is not a collection of random events but a carefully composed symphony, a design of breathtaking elegance.

Living in Harmony with the Code

To live with an awareness of sacred geometry and quantum patterns is to align oneself with the rhythms of existence. It is to see beauty not as an adornment but as the essence of reality, to recognize that every moment reflects the balance of the cosmos, that each thought and action contributes to the greater harmony.

This awareness transforms the ordinary into the extraordinary. A leaf becomes a cathedral of fractals, a wave a song of harmonics, a flower a meditation on unity. To honour these patterns is to live in harmony with the sacred blueprint of life, to move with the rhythms of creation, to become a participant in the cosmic dance.

The Eternal Symphony of Geometry

Sacred geometry and quantum mechanics reveal a universe that is structured yet fluid, finite yet infinite, random yet resonant. Beneath the surface of reality lies a code—a secret hymn that sings of unity, beauty, and love. To embrace this code is to step into the symphony of creation, to hear the music that flows through the stars, the waves, the spirals, the fractals.

This is the gift of sacred geometry, the revelation of quantum patterns: a reminder that we are not separate from the cosmos but an integral part of its harmony. Life itself is a shape, a pattern, a song—a sacred geometry etched into the infinite.

CHAPTER 15: THE COSMIC MIRROR – REFLECTIONS OF SELF IN THE QUANTUM WORLD

In the deep orchestration of existence, the cosmos holds a mirror—one that doesn't merely reflect the surface but peers into the depths of essence, revealing the intricate interplay of self and universe. Within this infinite looking glass, the quantum world whispers secrets of interconnection, entanglement, and the profound influence of awareness. It offers a revelation that the distinctions we perceive between observer and observed, self and other, are shadows cast by an overarching unity —a unity that vibrates within the very lattice of reality.

This cosmic mirror, unlike any terrestrial glass, refracts the nature of our being. It invites us to gaze into the fabric of existence and see not merely the external world but the internal—our fears, hopes, dreams, and the fundamental truths of our interconnectedness. The quantum realm does not merely describe particles and probabilities; it narrates the story of our own souls, reflecting the universality of our being and the vastness of our potential.

The Observer's Role: Weaving the Tapestry of Existence

In the realm of quantum mechanics, the act of observation is no passive endeavour. The wave function, containing all probabilities, collapses into a specific reality only when observed. Until then, it exists as potential, as possibility, as an unmanifested dream waiting for the touch of awareness. The observer, in this dance, becomes the catalyst, the one who ushers existence from the abstract to the tangible, the unformed to the formed.

To the mystic, this relationship mirrors the creative nature of consciousness. We are not merely spectators in life; we are co-creators, sculptors shaping the world through the chisel of thought, intention, and perception. Each moment, each choice, each act of attention is a brushstroke on the canvas of reality. To observe is to influence, and to influence is to create. The cosmic mirror teaches us that we are artists of our reality, participants in an unfolding masterpiece, threads in the universal loom weaving the tapestry of existence.

Entanglement: The Threads That Bind

Quantum entanglement reveals a startling truth: particles, once connected, remain so across distances that defy comprehension. The state of one particle instantaneously affects the state of the other, no matter how far apart they may be. This phenomenon transcends the confines of space and time, suggesting a reality where separation is illusory, where all things are intrinsically linked by invisible threads of unity.

In this web of entanglement, the mystic finds a reflection of the human soul. We are not isolated islands but interconnected beings, bound by threads of love, empathy, and shared existence. To harm another is to harm oneself; to uplift another is to uplift the whole. In the cosmic mirror of entanglement, we see the truth of our interconnectedness, a truth that dissolves the illusion of individuality and reveals the unity that lies at the heart of all things.

Entanglement whispers to us that the boundaries we place between ourselves and others are artificial. Each thought, each action ripples through the field of existence, touching lives, shaping realities, resonating through the invisible web that binds us all. To embrace this truth is to live with a deep sense of responsibility, knowing that the waves we create flow endlessly through the ocean of being.

The Duality of Particle and Wave: The Self in Flux

Particles in the quantum world exhibit a dual nature: they exist both as discrete points and as diffuse waves, embodying form and formlessness, the particular and the infinite. This paradoxical existence challenges the very notion of fixed identities, suggesting that reality —and by extension, the self—is fluid, multifaceted, and dynamic.

The mystic sees this duality as a profound metaphor for the human condition. We are both particle and wave, both a defined individuality and an expansive presence. At times, we are the wave, flowing, infinite, and unbounded. At others, we are the particle, distinct, anchored, and specific. Life is the dance between these states—a rhythmic oscillation between the particularities of our individual journeys and the universality of our shared essence.

To live as both particle and wave is to embrace the paradox of existence, to see oneself as both finite and infinite, both transient and eternal. It is to recognize that identity is not a rigid construct but a fluid expression of being a dance of becoming that flows with the rhythm of life.

The Cosmic Mirror: Reflections of the Inner Self in the Outer World

In the great quantum mirror, the external world reflects the internal. The patterns we observe in the universe—the waves, the entanglements, the probabilities—mirror the dynamics of our own consciousness. The mystic understands that the outer world is a projection of the inner, that the cosmos serves as a canvas upon which the self-paints its truths, fears, and aspirations.

To gaze into this cosmic mirror is to undertake a journey of self-discovery. Each interaction, each experience, each moment reflects the self, a window into the depths of one's being. The universe becomes a teacher, revealing our own nature through its patterns, inviting us to see ourselves in the stars, in the waves, in the infinite dance of existence.

This mirror reflects not only our light but also our shadow, showing us the aspects of ourselves that we often avoid. It reveals that to embrace the wholeness of existence, we must embrace the wholeness of ourselves, acknowledging both the beauty and the imperfections, the clarity and the mystery.

The Quantum Dance of Perception: Life as a Creative Act

In the quantum field, perception is creation. The observer's gaze determines the state of the observed, collapsing potential into reality. This principle reveals a profound truth: reality is not something that happens to us but something that happens through us. We are active participants in the unfolding of existence, creators shaping the world through the lens of our awareness.

The mystic understands this creative power as the essence of life. Each thought is a seed, each intention a wave, each action a force that shapes the world. To live with this awareness is to see life as a sacred act of creation, a dance in which every step, every movement, contributes to the symphony of existence. It is to recognize that we are not passive spectators but dynamic creators, beings who wield the power of consciousness to shape reality itself.

The Self as a Microcosm of the Universe

The cosmic mirror reveals a truth that mystics have long understood: the self is a microcosm of the universe, a reflection of the infinite within the finite. Just as the universe is vast and complex, so too is the self—a being of infinite potential, a soul that contains within it the essence of the cosmos.

To see oneself as a microcosm is to recognize that the universe is not separate from the self but is reflected within it. The patterns of the stars mirror the patterns of the mind, the rhythms of the cosmos echo the rhythms of the soul. In this vision, the self becomes a doorway to the infinite, a lens through which the universe contemplates itself, a reflection of the divine.

The Infinite Reflection: Living in Unity with the Whole

To embrace the cosmic mirror is to live in unity with the whole, to see oneself as part of the infinite, to recognize that life is a journey of reflection, a dance of interconnectedness, a symphony of being. This awareness invites us to live with love, to see each being as a reflection of ourselves, to honour the unity that binds all things.

The mystic understands that the cosmic mirror is not merely a metaphor but a reality, a truth that reveals the interconnectedness of existence, the oneness of life, the beauty of being. To live in harmony with this mirror is to live as a being of love, a soul who sees the divine in all things, a self who embraces both the light and the shadow, both the finite and the infinite.

The Endless Dance of Reflection

The cosmic mirror does not offer final answers but

infinite reflections. Each gaze into its depths reveals new truths, new dimensions, new facets of the self. To look into this mirror is to see not only the universe but oneself, to recognize that the journey of existence is a journey of self-discovery, a path that leads not outward but inward, into the heart of being.

This is the gift of the cosmic mirror—a reminder that we are not separate from the universe but are expressions of it, beings who reflect its light, who sing its song, who dance its dance. To live with this awareness is to embrace the infinite, to honour the interconnectedness of all things, to see life as a reflection of the soul and the soul as a reflection of the cosmos.

CHAPTER 16: THE THRESHOLD OF THE INFINITE – QUANTUM INFINITY AND SPIRITUAL AWAKENING

At the edge of comprehension, where the tendrils of human understanding graze the vast expanse of the unknown, a liminal space arises—a threshold where the finite mingles with the infinite, and the known dissolves into boundless possibility. Quantum physics beckons us toward this frontier, revealing glimpses of infinity nestled within the smallest particles, woven into the endless depths of space, and humming through the fabric of existence. Particles are not merely objects but probabilities, extending their reach across dimensions, whispering that what we perceive as finite is but a fragment of an unfathomable, infinite reality.

For the mystic, infinity is not an abstraction; it is a visceral encounter, a transcendent immersion in a boundless presence that defies articulation. It is an awakening to the eternal, a dissolving of personal boundaries into the unity of all things. To approach this threshold is to confront a paradoxical truth: the infinite is not distant but ever-present, not otherworldly but intricately intertwined with each moment of existence. Here, we delve into the intersection of quantum infinity and spiritual awakening—a confluence that unveils the infinite within and reshapes our understanding of life, self, and reality.

The Quantum Infinity: A Boundless Universe Within

Quantum mechanics unravels a universe that defies containment, a reality that brims with boundless possibilities. Particles, rather than static entities, exist as waves of probability, stretching beyond spatial and temporal constraints. Within this field of probabilities, no edges confine, no singular path dictates; everything is potential, flowing in an unending dance of creation and dissolution.

This quantum infinity hints that the physical world we perceive is a finite expression of an immeasurable whole—a kaleidoscope of energies extending into dimensions beyond our grasp. The finite is not diminished by this infinity but enriched, illuminated as a doorway to the limitless.

The mystic resonates with this vision of infinity,

perceiving the divine not as a distant deity but as an infinite presence imbued in all things. Each particle, each wave, each breath is a facet of this unending reality. To touch quantum infinity is to realize that existence is an unbounded field of potential—a living testament to the boundlessness of life itself.

The Threshold of Awareness: Dissolving the Boundaries of Self

Standing before the infinite is to approach the limits of self-awareness, where the familiar constructs of identity dissolve into an ocean of interconnected being. At this threshold, the mind's incessant narratives quiet, and what remains is presence—pure, vast, and unconfined. The boundaries that define "I" and "other" begin to blur, giving way to an expansive awareness that encompasses all things.

The mystic calls this encounter awakening, a moment when the veil of separateness is lifted, revealing the self not as an isolated entity but as an extension of the cosmos. In this space, individuality does not vanish but transforms, becoming a wave in the ocean of existence, a thread in the infinite tapestry of being.

Crossing this threshold is not an abandonment of self but an expansion—a realization that the self is a microcosm of the infinite, containing within it the essence of all creation. It is a journey into presence, where time collapses into the eternal now, and the soul rests in the boundless embrace of the infinite.

The Paradox of Infinity: The Boundless Within the Finite

Quantum physics reveals that infinity dwells within the finite. Even the smallest particle contains endless potential, an uncharted field of possibilities vibrating with life. This paradox—that the finite holds the infinite—is echoed in the mystical understanding of existence. Each individual, each moment, each fleeting thought contains the boundless within it, reflecting a truth far greater than its singular form.

The mystic embraces this paradox, understanding that life is a continuum where the ordinary becomes a portal to the extraordinary, where the ephemeral reveals the eternal. To live with this awareness is to see the sacred in the mundane, to recognize that each moment is a gateway to infinity, a passage that leads from the surface of existence to its immeasurable depths.

Infinity as Liberation: Transcending the Boundaries of Mind

To awaken to infinity is to experience a profound freedom—a liberation from the constraints of definition, limitation, and expectation. In the presence of the infinite, the rigid structures of the mind melt into fluid awareness, allowing the soul to expand beyond the confines of ego, culture, and identity. This is not an escape from reality but a fuller immersion into its essence, a recognition that true freedom lies in surrender to the vastness of being.

The mystic experiences this liberation as a state of pure

presence, a joyful union with the infinite that transforms fear into trust, division into unity, and isolation into communion. In this state, life ceases to be a struggle for control and becomes a dance with the divine, a flowing expression of the boundless love that permeates all existence.

The Divine Infinite: A Presence Beyond Boundaries

The divine, as experienced by the mystic, is not constrained by form, time, or space. It is infinite, a presence that flows through all things, uniting the material with the spiritual, the temporal with the eternal. To perceive the divine as infinite is to encounter a love that knows no conditions, a wisdom that transcends understanding, a reality that holds all contradictions in harmony.

This divine infinite is not separate from the world but is its very essence, imbuing every particle, every star, every soul with its boundless presence. To awaken to this infinite is to realize that the sacred is not elsewhere but here, now, within and around us—a truth that transforms existence into a holy communion with the divine.

The Infinite Field of Potential: A Journey of Becoming

In the quantum field, every moment brims with infinite possibilities—a canvas where each thought, intention, and action paints a unique reality. This field of potential reflects the essence of human existence: life as a dynamic process of becoming, an unfolding journey that is as boundless as the universe itself.

The mystic understands this potential as a call to co-create with the infinite, to live with intention, to shape reality not through force but through alignment with the flow of existence. To embrace the infinite field is to recognize that life is not a fixed script but an evolving masterpiece, a symphony where each note contributes to the whole.

Crossing the Threshold: Living in the Infinite Now

To step into infinity is to immerse oneself in the present moment, where the constraints of past and future dissolve, revealing the eternal now. In this state, life is not fragmented but whole, not fleeting but timeless. The mystic experiences this presence as a communion with the infinite—a space where the boundaries of time and self melt into the unity of being.

To live in the infinite now is to approach each moment with reverence, to see each breath as a reflection of the cosmos, each heartbeat as an echo of eternity. It is to dwell in the sacredness of life, to recognize that every instant holds the fullness of existence, the presence of the divine, the beauty of the infinite.

The Infinite Journey: Awakening to Boundlessness

The journey toward infinity is not a linear path but an eternal unfolding, a spiralling ascent that leads ever deeper into the mystery of being. It is a pilgrimage that reveals the infinite within the finite, the eternal within the moment, the divine within the self.

To embrace this journey is to live with an awareness of the infinite in all things—to see life as a dance of becoming, a celebration of existence, a reflection of the boundless love that flows through the cosmos. This is the gift of the threshold, the beauty of a life lived in communion with the infinite, the joy of a soul awakened to its unity with all that is.

The Infinite Embrace

To cross the threshold of the infinite is to surrender to the vastness of existence, to dwell in the unbounded presence of the divine, to live as a being who is both finite and infinite, both human and eternal. This is the wisdom of the mystic, the insight of the physicist, the calling of the soul—a recognition that life is not a destination but a journey into the heart of infinity, a journey that reveals the love, unity, and boundlessness of existence itself.

CHAPTER 17: QUANTUM ALCHEMY – THE TRANSFORMATIVE POWER OF INTENTION

In the quiet stillness where breath meets thought, there lies an ancient and unyielding force—a radiant energy veiled in simplicity: intention. Subtle as the whisper of dawn yet potent as a forge's fire, this force transforms potential into form, dreams into reality, and the unspoken into existence. It is the alchemist's crucible, the mystic's compass, and the unseen architect of the quantum field.

The ancient alchemists believed that the transformation of metals mirrored the evolution of the soul—that gold was not merely an element but the ultimate expression of spiritual refinement. Modern quantum physics echoes this truth, showing us that observation, choice, and focus actively shape the reality we inhabit. Intention, then, is more than a fleeting thought; it is a luminous force, a sacred fire capable of transmuting the mundane into the miraculous. To wield intention is to step into the role of creator, to engage in the divine alchemy where thought becomes form, and form becomes an expression of spirit.

The Sacred Flame: Intention as the Alchemical Catalyst

In the hands of the alchemist, the sacred flame does not merely melt metals; it illuminates the hidden pathways of transformation, purifying what is base and elevating it to what is noble. This flame is not merely external —it burns within the heart, fuelled by the clarity of purpose, the purity of focus, and the unyielding belief in possibility.

In the quantum realm, we find a parallel truth: reality is not static but a field of endless probabilities, waiting to be shaped by the focus of awareness. Intention, like the alchemist's flame, becomes the spark that ignites transformation, collapsing infinite potentials into a singular, realized outcome. Each thought, each choice, each act of will becomes a declaration to the universe, a ripple that cascades into creation, a wave that flows into being.

To carry this flame is to understand that transformation begins not with outer conditions but within. The philosopher's stone—the fabled key to alchemical mastery—is no mere object but the refinement of the self, the alignment of intention with the divine flow of existence. This is the essence of quantum alchemy: the recognition that reality is shaped not by accident but by the resonance of conscious intent.

The Quantum Field: A Canvas of Infinite Potential

Quantum mechanics unveils a reality unbounded by certainty—a universe of infinite potentials, where particles exist as waves of possibility, awaiting the focus of observation to collapse into form. This field of potential is not a void but a vast, living canvas, where every stroke of intention leaves its mark, where every thought is a colour, every action a brushstroke, every dream a spark of light.

To the mystic, this field is a sacred space, a dimension where the soul communes with creation, a place where the invisible becomes visible, where the unmanifest becomes manifest. Here, intention acts as the artist's hand, shaping the contours of experience, weaving threads of thought into the fabric of existence.

In this realm, every possibility coexists, every path unfolds simultaneously, and every moment holds the potential for transformation. The act of intention is the act of choosing—not from scarcity but from abundance, not from fear but from alignment. To live in harmony with this field is to embrace the power of co-creation, to see oneself as a participant in the great unfolding of the cosmos, a presence that dances with the infinite.

The Divine Forge: Intention as the Shaper of Reality

Just as the forge transforms raw ore into tempered steel, intention transforms the raw energy of possibility into the realized beauty of creation. But this forge requires more than heat; it demands focus, patience, and unwavering clarity. In the quantum world, intention becomes the artisan's hammer, shaping reality through the act of attention, bringing coherence to chaos, giving structure to the formless.

The mystic sees this forge as the heart's crucible, a place where the fire of love burns away the impurities of doubt, fear, and distraction. To live with intention is not merely to desire but to align—to harmonize thought, feeling, and action with the deeper currents of existence, to surrender ego to the greater wisdom of the soul. This is the alchemist's path: not a quest for control but an embrace of communion, a willingness to flow with the divine rhythm of creation.

The Dance of Creation: Intention in the Quantum Symphony

In the quantum field, particles do not move linearly but dance—a rhythmic interplay of waves and probabilities, a symphony of potentiality where each note influences the whole. Intention serves as the conductor of this symphony, directing the flow of energy, shaping the harmonics of experience, guiding the unfolding of reality.

This dance is not a performance for an audience but a participatory act, where each step, each movement, each gesture contributes to the greater whole. The mystic recognizes this dance as the essence of life, a journey of co-creation where each thought is a step, each feeling a rhythm, each action a melody. To dance with intention is to honour the interconnectedness of all things, to move in harmony with the cosmic flow, to participate fully in the unfolding of existence.

The Heart of Alchemy: Love as the Highest Intention

At the centre of every alchemical process lies the heart—the source of the purest intention, the wellspring of transformation. Love, the mystic teaches, is the highest form of intention, a force that transcends the boundaries of self and other, that heals, unites, and elevates. To act with love is to align with the divine, to bring harmony to the disharmony, to illuminate the shadows with the light of compassion.

In the quantum field, love becomes the ultimate force of coherence, a vibration that harmonizes the chaotic, a resonance that amplifies the beauty of the whole. To live with love as one's guiding intention is to transform not only the self but the world, to become a channel through which the divine flows, to bring the infinite into the finite, the sacred into the ordinary.

The Philosopher's Stone: Consciousness as the Key

The philosopher's stone, the alchemist's ultimate quest, is not a physical artifact but a symbol of inner mastery, a reflection of the soul's journey toward wholeness. In the quantum realm, this stone is consciousness itself—the awareness that transforms, the light that reveals, the presence that creates.

To hold this stone is to recognize that the self is both the alchemist and the crucible, both the creator and the creation. It is to see that life is not a series of fixed events but a dynamic interplay of intention, possibility, and realization. This is the true power of quantum alchemy: the ability to shape reality through the focus of consciousness, to transform the mundane into the miraculous, to awaken the divine within the self.

The Infinite Alchemy of Life

Life, when seen through the lens of quantum alchemy, is a continuous process of becoming, a journey where each moment holds the potential for transformation, each breath the opportunity for creation. Intention is the thread that weaves this journey, the flame that ignites it, the force that guides it. To live as an alchemist is to embrace life as a sacred process, to see each experience as a revelation, each challenge as an opportunity, each thought as a step toward the infinite.

This journey is not linear but spiralling, not finite but endless—a path that leads ever deeper into the mystery of being, ever closer to the heart of the divine. To walk this path is to live with purpose, to act with love, to create with clarity, to transform with intention.

Embracing the Alchemy of Intention

To embrace the alchemy of intention is to live as a co-creator with the universe, to see oneself as both the artist and the artwork, the dreamer and the dream. It is to recognize that each moment is a choice, each thought a spark, each action a brushstroke on the canvas of reality. This is the gift of quantum alchemy—a reminder that we are not passive observers but active participants in the dance of existence.

To live with this awareness is to awaken to the infinite power of intention, to see life as a sacred journey, a divine alchemy where the soul transforms the ordinary into the extraordinary, the finite into the infinite, the self into the divine. This is the journey of love, the dance of creation, the song of the soul—a journey that leads ever deeper into the heart of existence, a path that is, in truth, the journey of life itself.

CHAPTER 18: THE SILENCE BETWEEN – EMBRACING THE SPACE OF THE IN-BETWEEN

In the vast symphony of existence, it is not only the notes that create music, but the pauses between them. It is not only what is said that carries meaning, but also what is left unsaid—the breath between words, the moment before reply, the quietude before the storm. The universe, too, hums with these interstitial silences—spaces between becoming and being, between thought and action, between wave and collapse, between death and new life. They are not empty, these silences. They are full—dense with meaning, radiant with potential, and resonant with creation.

In the language of the mystic and the quantum physicist alike, the in-between is not absence—it is presence unspoken. It is a womb of creation. It is the seed not yet sprouted, the dream not yet dreamt, the universe not yet breathed into being. Between particle and wave, between question and answer, between longing and fulfillment—there, in that sliver of pause, lies the most powerful magic of all.

This chapter is an ode to that sacred pause. The silence between. The threshold space where all that is and all that will ever be waits—quietly, infinitely, patiently—for the right breath, the right gaze, the right moment of collapse.

1. The Quantum Threshold

Quantum physics, in its daring reconfiguration of reality, offers a strange and poetic revelation: that everything we see, everything we touch, everything we *believe* to be solid and real, is suspended in a delicate dance of probabilities until observed. Until an eye falls upon it—until consciousness engages—it exists as a potential, not a fact.

The wave function, Ψ, is the mathematical embodiment of this suspended reality. It does not describe what *is*—but what *could be*. Infinite outcomes coexist, overlaying one another in silent anticipation. This shimmering cloud of potentials, known as **superposition**, remains unresolved until observed. And in that act of observation, the wave collapses—choosing a form, a state, a here, a now.

But what of that space *before* the collapse? What of that trembling moment when all possibilities are still alive, when nothing has yet become, but everything *could*?

That, beloved reader, is the sacred in-between.

It is the space of becoming—the realm where the future leans against the present, whispering its many faces, awaiting the call of awareness to give it shape. It is neither past nor future, not a state nor a moment. It is liminality itself—a suspension of time, a pause in the architecture of the cosmos, where the eternal dwells in exquisite stillness.

2. The Mystic's Silence

Across time and tradition, mystics have known this space intimately. They have entered the space between thoughts, the void between breaths, the silence that follows a chant, and called it holy.

In **Tibetan Buddhism**, there is the concept of the **bardo** —a transitional space between death and rebirth. But bardo is not confined to death. There are bardos in every moment of our lives: between inhale and exhale, between waking and sleeping, between sorrow and joy. Each transition, each pause, is a bardo—a moment of infinite pliability, where the soul is unshaped, fluid, unanchored.

In **Hindu Vedanta**, the yogi speaks of **turiya**—the fourth state of consciousness, beyond waking, dreaming, and deep sleep. Turiya is the silent witness, the pure awareness that is present in the spaces between states. It is the stillness that underlies all movement, the backdrop upon which the play of life unfolds.

In **Christian mysticism**, the desert fathers went into silence not as escape, but as communion. They knew that the Word of God was heard most clearly in the hush, in the stillness where the self falls away and something vast arises.

And in **Sufi poetry**, Rumi writes:

"Silence is the language of God,
All else is poor translation."

These mystics, in their own tongues, describe the in-between as a sacred portal—a space where we meet not only the divine, but our truest selves, stripped of noise, of striving, of identity.

3. The Language of Liminality

Liminality—the state of being on a threshold—is not comfortable. It is not certainty. It is not clarity. It is mist and metaphor, ambiguity and ache. It is the caterpillar dissolved in its chrysalis, no longer what it was, not yet what it will become.

In the human experience, liminality appears in countless forms:

- The moment after diagnosis, but before healing or decay.
- The night before a wedding.
- The space between a goodbye and a hello.
- The years between leaving faith and finding something deeper.
- The breath you hold after "I love you" and before you hear it back.

These spaces are fragile and holy. They are the unshaped clay, the fertile soil of transformation. And they are often feared—because they cannot be controlled. They ask us to surrender. To dwell in *unknowing*. To listen to what can't be heard. To see what isn't shown. To feel what has not yet formed.

But in these spaces—these silences—we are rewritten. We are undone and remade. The old skin peels. The old roles dissolve. And in the quiet, we hear the whisper: *You are more than you thought. You are not who you were. You are not yet who you will be.*

You are infinite potential in pause.

4. The Alchemy of the Pause

Between cause and effect, between action and result, lies an alchemical field—a place where reality is still malleable, still responsive to the hand of the inner sculptor. Ancient mystics knew this as the realm of prayer. Not prayer as begging or barter, but prayer as presence—as becoming attuned to the hum of the cosmos and aligning intention with the deeper rhythms of creation.

Modern quantum theory, strange as it may seem, hints at something not dissimilar. The **Heisenberg Uncertainty Principle** tells us that we cannot know both the position and the momentum of a particle with absolute certainty. The more precisely we know one, the less we know the other. This uncertainty is not a limitation of measurement—it is woven into the very fabric of being.

Reality, then, is inherently indeterminate until interacted with. It is not locked down, but participatory. In this strange world of the small, the observer does not merely watch; the observer helps to create. Not by force. Not by logic. But by being there. By observing. By attending. The silence between cause and effect becomes the most powerful space of all—where consciousness participates in calling reality into form.

Mystics would nod at this. For they have long taught that presence itself is transformative. That awareness, when sustained and surrendered, can shift even the most unyielding aspects of life. The miracle, they say, is not in what happens, but in the space where it could. The silence before the miracle is the real miracle.

5. The In-Between as Mirror of the Self

We often think of life as made up of moments—birthdays, marriages, achievements, losses. But if you trace your truest transformations, you may find that the moments themselves were less important than the spaces *between* them. Those inarticulate, often painful seasons where you were no longer who you used to be, and not yet who you were becoming.

These are the wilderness years, the gestation phases, the nights when sleep won't come and the stars seem to whisper questions you cannot answer. These are the liminal corridors where identity unravels—when the masks fall away, and the naked self, shivering and luminous, looks into the mirror of silence and asks, *Who am I, really, without all this?*

In these interstitial spaces, something strange happens. The old narratives lose their grip. The ego, deprived of its usual roles and rhythms, becomes quiet. And in that silence, something ancient stirs.

Not knowledge. Not belief. But *awareness.*

You realize that you are not the roles you played. Not the fears you fed. Not the hopes you chased. You are something quieter. Something deeper. You are the awareness *between* all that. The consciousness that watches the rise and fall of thought, the breath between heartbreak and healing, the presence that persists no matter what name the world gives you.

This is the mystic's path—to become not a person who knows more, but a soul who *rests* more deeply in the unknown.

6. Death and the Great In-Between

Perhaps the most profound silence of all is death—not merely the event, but the space it opens. Death is not the end, say the mystics. It is a transition. A passage. A release from form into formlessness. A return to source.

In quantum physics, particles are born and annihilated in an endless dance. Energy never disappears—it changes state. It moves from expression to potential, from manifestation to silence. What is death if not the ultimate interstice? The pause between the known world and the great mystery beyond?

Near-death experiences, across cultures, describe a similar threshold—a tunnel, a light, a space where time ceases, where identity dissolves, where peace envelops. These may be glimpses into that same quantum field of possibility—a place where consciousness, unmoored from the body, becomes fluid, infinite, whole.

But we need not wait until the end to touch that space.

Each night, in sleep, we enter the in-between. Each meditation. Each moment of stillness. Even grief—so painful, so aching—is a form of liminality. It ungrounds us from the world we knew, opens a raw space inside us, and beckons us toward depth. Toward truth. Toward presence.

What if the "end" is not a door closed, but a veil parted?

What if the silence that follows death is not absence, but an invitation?

7. The Silence as Source

The in-between is not just a moment. It is a dimension. A layer of reality as real as the seen. Mystical traditions teach that the world arises *from* silence—that all sound, all matter, all movement is a ripple in the stillness of the void.

In the Hindu Upanishads, there is a mantra:

"Om. That is full. This is full. From fullness comes fullness.
When fullness is taken from fullness, fullness still remains."

This speaks of a reality that is not diminished by giving, not lessened by expression. Silence, too, is such. When the universe speaks, it does not deplete the silence—it arises from it. The silence is the source. The ever-fertile field. The quantum vacuum, teeming with zero-point energy, shimmering with unrealized forms.

Physicists now believe the so-called "vacuum" of space is not empty at all. It is alive. Foaming with virtual particles, popping into and out of being, lending energy to reality in unseen ways. The vacuum is full, not void. Like the mystical Silence—it holds all that is, and all that could be.

To enter this silence is to return to source. To rest in the field where all things begin. To commune with the place where consciousness dreams the cosmos into form.

8. Embracing the Interstice: A Life Between

What would it mean to live *from* this place? To make a life not on the extremes, but in the gentle valley between them? To move not from certainty, but curiosity? Not from answers, but questions?

Such a life would not be linear. It would be spiral. It would unfold in spirals of becoming—each turn a new pause, each breath a new creation. It would be less about achievement and more about attunement—listening for the music behind the silence, the light behind the form.

It would be a life of presence.

You would stop rushing to resolve. You would let the unanswered stay a little longer. You would find beauty in the unfinished. Grace in the pause. Love in the liminal.

And you would realize: the silence between is not a waiting room. It is the temple itself.

9. Becoming Silence: The Deep Listening

To dwell in the in-between is to become not merely a witness of silence, but a vessel for it. This is not a passive act. It is the most sacred kind of listening—not with ears, but with being. A kind of receptivity that requires the ego to grow still, the mind to ungrasp, the body to soften into the cradle of existence itself.

Mystics speak of **deep listening**—a surrender to the pulse of the cosmos. In Sufism, it is called *sama*, the divine hearing, where the seeker becomes attuned to the subtle vibrations of the universe, where even silence becomes a voice of God.

In the Kabbalah, silence is seen not as lack, but as a mode of divine communication—*shevirat ha-kelim*, the shattering of vessels, where divine light breaks through form and radiates through the spaces between things. God is known not just in speech, but in the absence of speech—in the spaces left untouched.

And in Zen Buddhism, the koan becomes the tool of silence—a question without answer, a riddle that breaks the mind's need for resolution, and returns the seeker to the immediacy of presence. *What is the sound of one hand clapping?* The question is not for the intellect—it is a doorway into the place where the questioner dissolves.

The in-between, then, is not just an event or location—it is a quality of consciousness. A way of *being* in the world that is porous, curious, soft. A kind of sacred receptivity. And in that deep listening, something strange and luminous begins to occur: you begin to hear the world differently. The spaces between words begin to shimmer. The pauses in a conversation become as meaningful as the speech. The moment between a touch and its response becomes holy.

You begin to hear not just with ears, but with the soul.

10. The Space Between People

Liminality is not just personal or cosmic—it is relational. There exists, between every two people, a sacred space. A third presence. A field. It is not made of words or gestures. It is made of *attention*. Of *intention*. Of the willingness to pause and let the other *be*—not as an object to understand or fix, but as a mystery to behold.

In quantum mechanics, entanglement tells us that two particles, once connected, remain linked across time and space. Their states are not separate—they are correlated. What happens to one affects the other, instantly, across distance.

So it is with souls.

We meet someone. We exchange a glance, a word, a gesture. And something invisible is born between us. A field. A subtle thread. A silence charged with meaning. And even when we part, that thread may remain, shimmering, alive.

What if love is not just emotion, but entanglement? What if friendship, family, devotion are quantum threads woven between beings—linking us in ways science is only beginning to grasp?

The space between people is a creative space. It is where empathy blooms, where trust is built, where healing begins. But only if we honour it. Only if we learn to pause. To enter the silence between words and offer the gift of presence.

Every argument can be softened by a pause.

Every misunderstanding can be reframed by a breath.

Every connection deepened by silence.

The in-between is where we remember: the other is not separate. The other is not object. The other is a reflection of the same light that dwells in us.

11. Technology and the War on Silence

We live in an age that fears silence. We fill every pause with music, with screens, with endless scrolling. We are addicted to motion, to noise, to input. We have forgotten the value of stillness, the sanctity of not-knowing, the artistry of the pause.

Technology has collapsed many in-betweens. Messages are instant. Answers are a click away. We do not wait. We do not wander. We do not sit at the edge of the unknown and tremble with reverence.

But the soul still remembers.

It aches for the quiet spaces. It longs for the walks without purpose. The stare into the void. The long gaze at the stars. The gentle waiting that feels like prayer. The embrace of mystery.

To reclaim the in-between is a radical act.

It is to say: I will not be ruled by urgency.

I will not chase every answer.

I will not flee from silence.

I will make space in my life—for wandering, for pausing, for the sacred unknown.

12. The Cosmic Implication: Existence as Interstice

At the grandest scale, the cosmos itself may be a kind of in-between. A shimmering tapestry suspended between non-being and being. Between the infinite potential of the quantum vacuum and the embodied unfolding of galaxies.

Astrophysics tells us that most of the universe is

dark. **Dark matter. Dark energy.** Unseen. Unknown. Unmeasured. We know it exists by what it *affects*—by how it shapes the dance of stars. But we do not see it. We do not touch it. It is, like silence, a hidden force that holds all things together.

Is it absence? Or is it a deeper presence?

Is the cosmos made of things? Or of *between-things*?

Mystics say: all is One. Quantum physicists say: all is entangled. Between these statements lies a shimmering unity. A mystery vast and tender. A field of interconnected becoming.

In this vision, *you* are not a separate self-moving through a solid world. You are a ripple in a relational field. A momentary knot in the quantum net. You are *between*—between past and future, self and other, wave and particle.

You are silence becoming speech.

You are potential becoming form.

You are the breath before the word.

You are the pause that sings.

13. A Life Shaped by Silence

If all this is true—if the in-between is not a gap to fill, but a gift to receive—then how might we live differently?

We might begin to *honour* the silence. To seek it, even.

We might create spaces in our day not for achievement, but for listening.

We might hold our grief with more patience, knowing it is not a flaw to be fixed but a passage to be walked.

We might release the need for constant knowing and let the unanswered live in us like seeds.

We might stop rushing. Start listening. Start noticing the spaces *between* things.

We might begin to shape our lives not as stories to control, but as dances to be danced with the rhythm of breath and stillness.

We might become the kind of people who are comfortable in pauses. Who are present in transitions. Who are soft in the face of the unknown.

We might become vessels of silence.

And in doing so, we might find ourselves again—not as fixed beings, but as flowing, spacious, luminous expressions of the great silence that births all things.

14. A Benediction of the In-Between

To you who walk the liminal path...

May the silence hold you gently.

May the space between answers become a sanctuary, not a prison.

May the pause between heartbeats remind you of your aliveness.

May the gap between thoughts become a gateway to presence.

May the mystery that waits in the interstices be your teacher, your guide, your friend.

May you learn to live not on the edge of silence, but *within* it.

Not fearing the unknown, but dancing with it.

Not rushing to resolve, but revering the unfolding.

May you become, like the quantum field itself, a vessel of potential.

A silence full of stars.

A breath of the infinite.

A sacred in-between.

CHAPTER 19: THE SOUL'S SPIN – MYSTICISM AND QUANTUM SPIN

In the realm of the quantum, particles exhibit a peculiar, intrinsic form of rotation known as spin. Unlike classical spinning objects, quantum particles' spin does not involve a literal rotation in space but is instead a fundamental property, a mysterious force that defines their orientation, their behaviour, their dance within the fabric of existence. This spin is dynamic, alive, a constant motion that pulses through each particle, hinting at an invisible momentum, a force of becoming that defies the stillness of definition.

So, too, does the soul spin, caught in the currents of life, drawn to growth and change, moving perpetually in the dance of existence. The mystic feels this spin as the soul's journey—a ceaseless unfolding of self, a path that moves in cycles, spirals, a rhythm that draws the self inward and outward in a dance of spiritual transformation. Here, we explore the concept of quantum spin as a metaphor for the soul's journey, a reminder that life is not a linear path but a dynamic dance, a motion that reflects the eternal becoming of a soul in search of truth.

The Spin of Particles: A Dance Beyond Definition

In the quantum world, spin is a property that defies ordinary understanding, a state that does not conform to the dimensions of classical rotation but instead represents a fundamental aspect of a particle's existence. Electrons, protons, and neutrons each possess spin, a motion that influences their interactions, their orientations, their magnetic moments. This spin is not simply a feature; it is a defining aspect, a mark of identity, a symbol of the particle's inner nature, an energy that speaks to the essence of its being.

For the mystic, the soul, too, possesses a spin, a dynamic quality that defines its journey, a motion that flows from its core, a rhythm that guides its path. This spin is not a movement through space but a turning of awareness, a flow of energy that moves the self through cycles of understanding, phases of growth, moments of transformation. To feel the soul's spin is to recognize that life is not static but alive, a journey that flows in circles and spirals, a path that leads not to a final end but to an ever-deepening awareness of the infinite.

The Dance of Duality: Spin as a Balance of Opposites

Quantum spin embodies a fundamental duality—particles can spin "up" or "down," orientations that represent opposing states, forces that are distinct yet interconnected, polarities that shape the behaviour of particles. This duality is not a separation but a balance, a harmony of opposites that reveals the interconnectedness of all things, a reflection of the cosmic dance of light and shadow, yin and yang, positive and negative, self and other.

In the journey of the soul, there is a similar dance of opposites, a balance between light and darkness, joy and sorrow, growth and surrender. The mystic understands that to grow is not to escape duality but to embrace it, to see that the soul's journey involves both ascent and descent, both expansion and contraction, both becoming and letting go. To live with an awareness of this duality is to recognize that each phase, each cycle, each spin of the soul is part of a greater whole, a dance that reveals the unity of all things, a journey that holds both the light and the shadow within its embrace.

The Spin of Consciousness: Movement as a Path to Awakening

In the quantum realm, spin is a source of energy, a motion that influences the interactions of particles, a property that defines the nature of matter itself. Similarly, in the spiritual journey, the soul's spin is a force that moves consciousness, a motion that awakens awareness, a rhythm that leads the self through cycles of understanding, phases of transformation, moments of awakening. This spin is not a random motion but a purposeful journey, a path that leads the self to deeper truths, to higher understanding, to a fuller expression of its own essence.

To feel the spin of consciousness is to experience life as a journey of awakening, a process of becoming, a dance that leads the self to ever-deeper layers of being, ever-higher states of awareness. The mystic understands this journey as a spiral, a path that moves both inward and outward, a rhythm that reveals the inner truth of the self, the essence of the soul, the divine within the depths of being. To embrace this spin is to recognize that each moment, each experience, each thought is a step in the journey of awakening, a step that leads the self-closer to the infinite, to the divine, to the truth of who we are.

The Quantum Twist: Entangling the Soul with the Infinite

In the quantum world, particles are not isolated entities but are often entangled, their spins intertwined in such a way that their states are inseparable, their fates bound, their movements connected across distances, across time. This entanglement is a mystery that defies explanation, a connection that transcends space, a bond that reveals the unity of all things, a dance that connects each particle to the whole.

The mystic feels this entanglement as the essence of the soul's journey, a recognition that the self is not separate but is woven into the fabric of existence, a part of the cosmic dance, a soul whose spin is connected to the divine, to the universe, to all beings. To live with an awareness of this entanglement is to see that each thought, each action, each intention is not isolated but resonates through the field of consciousness, touching all things, shaping the journey of the soul, revealing the interconnectedness of life. This is the gift of the soul's spin—a reminder that we are part of the whole, beings who are woven into the dance of existence, souls whose journeys are intertwined with the infinite.

The Spiral of Growth: The Soul's Journey Through Cycles of Transformation

In the journey of the soul, growth is not a straight line but a spiral, a path that returns to familiar places, familiar challenges, familiar truths, each time at a deeper level, each time with a greater understanding, a fuller awareness, a deeper love. This spiral is the essence of the soul's spin, a journey that moves in cycles, a rhythm that leads the self through phases of becoming, a dance that reveals the beauty of change, the wisdom of impermanence, the joy of transformation.

The mystic understands this spiral as the path of growth, a journey that leads the self through layers of awareness, levels of understanding, states of consciousness. To embrace this spiral is to recognize that life is not a single journey but a series of cycles, a series of spins that lead the soul through the dance of existence, a journey that reveals the infinite within the finite, the boundless within the boundaries of time and space. This is the beauty of the soul's spin—a reminder that life is not static but alive, a journey of becoming, a process of continual awakening, a path that leads ever deeper into the heart of the infinite.

The Inner Rotation: Finding Stillness Within the Spin
In the spinning dance of particles, there is a paradox—a stillness within motion, a quiet within the energy, a silence at the centre of the spin. This stillness is not an absence but a presence, a calm that lies at the heart of movement, a peace that holds the energy within its embrace. The mystic feels this stillness as the essence of the soul, a quiet that lies beneath the surface of thought, a presence that transcends the motion of life, a silence that reveals the true nature of the self.

To find this stillness is to experience life from the centre of the spin, to feel the movement without being moved, to embrace the rhythm without losing oneself in the dance. In this stillness, there is a clarity, a wisdom, a peace that transcends the cycles of change, a presence that reveals the infinite within the self, the divine within the silence. This is the essence of the soul's spin—a reminder that life is not a series of random events but a dance of awareness, a journey of awakening, a path that leads the self to its own centre, a place of stillness within the spin.

The Soul's Spin as a Reflection of the Divine
The spin of the soul is not separate from the spin of the universe, a reflection of the divine dance, a rhythm that resonates with the pulse of the cosmos, a motion that reveals the unity of all things. The mystic understands that the soul's journey is not an isolated path but a part of the whole, a dance that is woven into the fabric of existence, a journey that reflects the essence of the divine.

To live with an awareness of the soul's spin is to see each moment as a part of the greater dance, to recognize that each thought, each feeling, each action is a step in the journey of the whole, a note in the symphony of life, a wave in the ocean of consciousness. This is the gift of the soul's spin—a reminder that we are not separate from the universe but are part of it, beings who resonate with the essence of existence, souls who dance to the rhythm of the divine.

Embracing the Spin: Living in Harmony with the Journey of the Soul
To embrace the soul's spin is to live in harmony with the journey of life, to recognize that each phase, each cycle, each moment is a part of the whole, a step in the journey of becoming, a reflection of the infinite within the finite. This awareness invites us to live with love, to embrace both the light and the shadow, to see each experience as

a part of the dance, each challenge as a step in the path, each joy as a note in the song of existence.

The mystic understands this journey as a path of awakening, a process that reveals the beauty of the soul, the truth of the self, the love that flows through all things. To live in harmony with the soul's spin is to see life as a journey of creation, a dance that leads the self to ever-deeper levels of awareness, a path that reveals the infinite within the heart of the finite.

The Infinite Dance of the Soul

And so, we return to the soul's spin, to the recognition that life is not a static state but a dynamic journey, a process of continual awakening, a dance that reveals the beauty of existence, the wonder of being, the love that flows through all life. This is the gift of the soul's spin—a reminder that the self is not bound by time or space, that consciousness is not confined by form, that life is a journey of becoming, a path that leads ever deeper into the heart of the infinite.

To embrace the soul's spin is to live with an awareness of the infinite, to see each moment as a step in the dance, each breath as a beat in the rhythm of life, each thought as a wave in the ocean of consciousness. This is the path of the mystic, the journey of the soul, a journey that leads not to answers but to love, not to certainty but to wonder, not to knowledge but to the infinite.

Embracing the Soul's Spin

To embrace the soul's spin is to live as a being of love, a soul who sees the sacred in each moment, who feels the presence of the divine in each thought, each action, each motion. This is the wisdom of the mystic, the vision of the philosopher, the insight of the physicist—a recognition that life is a journey of creation, a path that leads us ever deeper into the heart of existence, a journey that reveals the beauty of the soul, the truth of the self, the love that flows through all things.

This is the gift of the soul's spin, the beauty of a life lived in harmony with the field of becoming, the joy of a journey that leads us ever deeper into the heart of the infinite, a journey that is, in truth, the journey of love.

CHAPTER 20: THE MYSTIC'S EYE – PERCEIVING BEYOND THE VEIL

Beneath the surface of the known, where reason falters and silence hums, lies a veil—a delicate gossamer that shrouds the deeper truths of existence. It is not a veil to obscure but to invite, not to conceal but to beckon. Beyond it stretches the infinite, the sublime, the sacred geometry of all that is and will ever be. Mystics call it the threshold of the divine; scientists name it the edge of comprehension. Both are seekers, kindred in their longing to pierce the illusion and glimpse the unity that breathes beneath the multiplicity of forms.

To perceive beyond this veil is not to escape reality but to immerse oneself in its essence, to allow the world's hidden symphony to resound within the soul. It is a journey of unseeing and seeing anew, where the mystic's eye and the scientist's vision converge in the luminous depths of the unutterable.

The Veil: A Threshold of Mystery

Human senses, intricate though they are, grasp only fragments of existence, interpreting the vast symphony of being through narrow keys. Vision is confined to light's narrow spectrum; touch is limited to what is solid, taste and smell to what dissolves. Yet, reality dances far beyond these boundaries, moving in wavelengths imperceptible, resonating in dimensions that eyes and hands cannot reach.

This veil of perception is not a wall but a threshold, a liminal space that both reveals and conceals. It whispers of vastness, teasing the edges of curiosity, inviting the seeker to look beyond the immediate and the apparent. To the mystic, the veil is a paradoxical gift—a veil to be transcended and yet revered, for its presence invites the quest for insight. For the scientist, it is a frontier, an uncharted land where instruments and intellect must expand to meet the challenges of the infinite. Both understand that the veil is not a barrier but a mirror, reflecting back the depth of our yearning to know.

The Mystic's Eye: Seeing with the Inner Light

For the mystic, to see beyond the veil is not to strain outward but to turn inward. The mystic's eye is an aperture of the soul, an inner lens polished by stillness, love, and the letting go of illusion. This eye does not see with light but with presence, perceiving not objects but essences, not surfaces but depths.

To open this eye is to awaken an awareness that

transcends duality, to perceive the infinite woven within the finite, the eternal held within the fleeting. The mystic's vision is one of interbeing—seeing not a world of isolated forms but a single, boundless unity. A river is not merely flowing water but the pulse of life; a tree is not merely wood and leaf but a living prayer, a hymn rooted in earth and stretched to sky.

The mystic's eye beholds the world as sacred, each fragment a fractal of the whole, each moment imbued with the numinous. It is a seeing that is not of the eyes but of the heart, a perception that does not dissect but embraces, that does not grasp but surrenders.

The Scientist's Vision: Peering into the Fabric of Existence

Where the mystic turns inward, the scientist gazes outward, armed with instruments that extend the senses into realms unseen. Telescopes peer into the abyss of galaxies; microscopes unveil the teeming cosmos within a drop of water. The scientist's vision dissects the atom, untangles the DNA spiral, measures the quantum pulse. Yet, the deeper they look, the more they encounter mystery.

Quantum mechanics reveals a reality that defies solidity, a universe composed not of things but of waves, probabilities, and relationships. Electrons do not orbit like planets but appear and disappear, flickering like fireflies. Particles entangle, communicating instantly across vast distances as if whispering across dimensions.

The scientist's vision, like the mystic's eye, unveils interconnectedness. It reveals that the observer and the observed are bound in a dance, that perception itself shapes reality. The closer the scientist approaches the veil, the more their findings resonate with the mystic's ancient knowing: reality is not a machine but a miracle, not a sum of parts but a seamless whole.

The Threshold of Mystery: Facing the Infinite

At the edge of understanding, both the mystic and the scientist encounter a shared truth: the ultimate nature of existence cannot be captured by thought alone. Knowledge expands yet never reaches completion; insight deepens yet always meets the unknown. This threshold is the birthplace of wonder, a space where humility replaces certainty, and awe dissolves the boundaries of ego.

For the mystic, this threshold is a doorway to the divine, an encounter with the infinite that silences all questions. It is not a space of answers but of communion, where the soul bows before the vastness of being, surrendering its illusions of control.

For the scientist, it is the frontier of exploration, a reminder that every discovery reveals further mysteries, that the universe is not a puzzle to be solved but a story to be continually unfolded. The threshold of mystery is where the known meets the unknowable, where the finite glimpses the infinite, where the quest itself becomes the destination.

The Mystic's Compass: Intuition and Inner Knowing

The mystic's journey beyond the veil is guided by

intuition—a deep, inner knowing that arises not from logic but from resonance. Intuition is the whisper of the infinite within the finite, the soul's compass pointing toward truth. It is not a substitute for reason but a complement, a vision that sees with the heart what the mind cannot grasp.

This inner knowing perceives reality not as an object to be analysed but as a presence to be felt. It recognizes the divine not in distant heavens but in the pulse of life, in the breath of being, in the silence that underlies all sound. Intuition is the mystic's bridge across the veil, a way of seeing that is not about answers but about attunement, not about certainty but about communion.

The Vision of Love: Perceiving the Sacred

The mystic's eye is not merely a lens of insight but a window of love. To see beyond the veil is to see with compassion, with a recognition of the sacred in all things. Love is the true light of perception, the force that illuminates the unity beneath the diversity, the essence beneath the form.

To perceive with love is to dissolve the barriers between self and other, to see each being as a reflection of the whole, each moment as a manifestation of the divine. Love is not merely an emotion but a state of awareness, a way of seeing that transforms perception into connection, observation into participation, sight into communion.

Beyond the Veil: The Infinite Within

To pierce the veil is not to escape the world but to enter it fully, to see the infinite within the finite, the eternal within the fleeting. It is to recognize that the veil itself is an illusion, a construct of perception, a boundary that exists only in the mind. Beyond it lies not a separate reality but a deeper dimension of the same reality—a realm where everything is connected, where every particle reflects the whole, where every soul resonates with the infinite.

This is the mystic's ultimate vision: to see that there is no veil, no separation, no duality. The world itself is sacred, and to live is to participate in the divine dance, to embrace the infinite within the everyday, to see the divine reflected in all things.

Living Beyond the Veil

To live with the mystic's eye is to see the world as it truly is—not as a collection of objects but as a communion of souls, not as a series of events but as a continuous unfolding of being. It is to walk with awareness, to act with love, to perceive with reverence.

The mystic's eye is not reserved for the few but is a gift that lies within each of us, waiting to awaken. It is the vision that sees beyond appearances, the insight that recognizes the unity of all things, the love that perceives the divine in every moment.

This is the gift of the mystic's eye: the ability to see the infinite within the finite, the sacred within the ordinary, the boundless within the bounded. It is the vision that pierces the veil, the awareness that embraces the whole, the perception that transforms life into a journey of wonder, a path of love, a dance with the divine.

CHAPTER 21: THE ART OF COLLAPSE – CREATION THROUGH COLLAPSE OF THE WAVE FUNCTION

There are moments in life so subtle that they almost go unnoticed. A decision made in a quiet room. A glance shared across a crowded space. A breath held a second longer than usual. These are the moments that shape everything — the imperceptible thresholds between what might have been and what comes to be. In the language of quantum physics, these moments mirror the phenomenon known as the *collapse of the wave function*. In simpler terms, it is the precise instant when a particle, once existing in a cloud of possibilities, assumes a definite state. Reality, quite literally, takes form — not gradually, not predictably, but all at once.

And though it is studied in laboratories with electrons

and photons, this moment of collapse is not confined to the world of subatomic particles. It lives within us. It surrounds us. It defines the human experience. In many ways, it *is* the human experience — a perpetual act of observing, choosing, collapsing the formless into form.

To understand this concept, one must begin with the essence of the wave function. In quantum theory, particles such as electrons do not exist in a single, definite place or state until they are measured. Instead, they exist in superposition — a suspended animation of potential outcomes, each with its own probability. The wave function is a mathematical description of all these possibilities. It is, in a sense, the whisper of what could be.

The mystery, and indeed the poetry, of the wave function lies in its fragility. It endures only as long as it is left undisturbed. The moment an observation is made, the wave collapses. The particle chooses a position, a velocity, a state. The infinite becomes finite. Possibility becomes history.

For physicists, this collapse poses a profound question: what triggers it? Is it the act of measurement, a physical interaction, or something more elusive — the presence of a conscious observer? Some, like Eugene Wigner, have suggested that consciousness plays an integral role. That reality itself may be, at least in part, shaped by awareness. The observer does not simply witness the world but participates in its unfolding.

To spiritual seekers, this is no surprise. The ancient mystics and seers spoke of this long before quantum equations gave it modern form. They understood the world not as a fixed structure, but as a field of infinite potential. They knew that reality bent toward intention, that creation arose not from force, but from focus.

A yogi sitting in meditation is not idle; he is collapsing his internal waves — refining intention, aligning thought, energy, and awareness into coherence. The mystic's prayer, the artist's brushstroke, the moment a mother first names her child — all are acts of collapse. All are decisions that take root in the unmanifest and become real.

In this light, the quantum world and the spiritual world no longer seem disparate. They begin to echo each other. Physics describes the mechanics. Mysticism explores the meaning.

But what does this mean for us, as individuals, as conscious agents of creation? It means that each thought, each observation, is not passive. It is generative. It means that intention is not only a psychological event but a quantum one. The choices we make — especially those made with clarity and awareness — ripple out into the field, collapsing potential into form.

This understanding invites a new way of living. Imagine waking each morning not merely to react to the day but to collapse it. To enter into it not as a victim of circumstances but as a conscious participant in its design. What if we chose not just what to do, but *how* to see? What if our very perception shaped the path ahead?

The mystics have long practiced such living. The Sufi dervish, spinning into trance, collapses the self into union. The Zen monk, sweeping the temple steps in silent attention, collapses awareness into each movement. The desert hermit, seated alone beneath a canopy of stars, collapses the illusion of separation and touches the eternal.

Their lives are not escapes from the world but deeper entrances into it — through the quiet art of collapsing the field with presence, devotion, and humility.

Even in the most mundane moments, this art is available to us. The simple act of listening — truly listening — is a collapse. Instead of tuning out or projecting assumptions, we attend. We choose one thread of another's voice, one truth amid many. And in doing so, we draw forth the real. We say: *This is what I will bring into being.*

The world is overflowing with potential. But potential alone is not enough. It must be met. It must be held. It must be observed. Not just with the eyes, but with the heart.

And so, the collapse of the wave function becomes more than a scientific mystery. It becomes a spiritual metaphor — a reminder that each of us, in our own quiet way, is an artist of existence. That each choice, each thought, each act of faith or fear, is a brushstroke on the canvas of the cosmos.

The wave, shimmering with what could be, waits patiently for your gaze. For your yes. For your willingness to participate in the grand unfolding.

Reality, then, is not a sentence handed down from on high. It is a poem written in real time — by you, by me, by the interplay of infinite potentials and the humble, trembling collapse of each one.

There is a certain power in knowing that reality does not simply unfold from some impersonal law, but through a living conversation between the world and the one who perceives it. The observer, long thought to be separate from what is observed, now finds themselves at the centre of the mystery. In both physics and mysticism, the veil separating the seer from the seen has grown thin — so thin that it dissolves entirely, if we are willing to see clearly.

To collapse a wave is to end ambiguity. In the quantum realm, it is the act that ends superposition — the simultaneous being of multiple, contradictory states. In human life, too, we often live in states of superposition: moments where we are undecided, where our future is suspended in possibility. When we finally act, something solid forms. One life continues, another quietly ends.

And often, we fear this collapse. Why? Because it requires commitment. Because it means that a hundred possible selves must be let go so one can be born. Because once the wave collapses, the field is no longer infinite — it is particular. It is now. It is here. And here, the illusions of safety we once found in abstraction vanish.

But there is also profound beauty in this. To choose — truly choose — is to engage with life in its most elemental form. It is to look the unformed in the eye and say, "Yes, I will shape you. I will meet you. I will walk this path."

In this way, the collapse of the wave function is not just physics — it is courage.

It is the courage to live without guarantee.

It is the courage to create without knowing if the canvas

will hold the paint.

It is the courage to speak, to love, to let go.

Each of us has had experiences of this. A career path decided after long silence. A child named with trembling joy. A relationship entered not with certainty but with sincerity. These are all collapses — intentional, soul-led, mysterious. They mark the difference between drifting and devotion.

The ancients understood this too, though they did not speak in the language of quanta. In the Upanishads, the sages wrote of the **bindu** — the point at which all that is latent becomes manifest. It is the meeting point between Shiva and Shakti, between stillness and movement, between potential and expression.

Modern physicists describe something very similar when they speak of decoherence. Before measurement, a quantum system holds many possible outcomes. But once the environment interacts with it — once the context is set, the observer enters, the system is "seen" — it collapses. The probabilistic cloud resolves into form. This is decoherence: the loss of ambiguity. The translation of uncertainty into experience.

Yet, in both cases, whether through the bindu or through decoherence, the principle remains: intention shapes the field.

And so the question becomes: *What intentions are we collapsing into reality?*

Are we living unconsciously, letting our fears and habits collapse our days for us?

Or are we pausing, breathing, and choosing the thread we wish to walk?

This is the spiritual art of collapse — not a surrender to fate, but a collaboration with the unknown. It does not mean control. Control is brittle. What it means is alignment. Attunement. A willingness to meet the unmanifest not with demands, but with devotion.

In this way, reality becomes more than external events. It becomes the echo of attention. It becomes sacred mirror. It becomes prayer answered not in words, but in the unfolding of the moment.

And so, when we speak of manifestation — of calling something into being — we are not speaking of magic in the childish sense. We are speaking of the conscious collapse of potential. We are speaking of the practice of clarity, presence, and trust.

For collapse is not merely choice — it is choice in alignment with the deeper field. With the will that breathes through all things.

To live as a spiritual artist of collapse, then, is to become a student of stillness. It is to recognize the waves of possibility not as distractions, but as invitations. And it is to collapse not in haste, but with reverence.

There is a moment before every major act in life — whether sacred or ordinary — where time seems to slow, where the field pauses, where your soul holds its breath. That moment is the threshold. And what you bring into it — your belief, your fear, your clarity — determines what collapses on the other side.

The mystics speak of this in the language of silence. They say that true manifestation arises from the *pause* — that emptiness is not a lack, but a space where meaning is born.

Quantum theory agrees. Before a particle is measured, it is undefined. The silence before collapse is not absence. It is the most fertile presence of all.

Thus, the wise learn to linger in the pause.

They do not rush to fill the silence. They do not demand answers too soon. They know that the unmanifest must be honoured before it becomes form.

They know that what collapses reflects not just what they want, but who they are.

To the untrained eye, the collapse of a wave function might seem like a scientific abstraction. It happens in cold chambers, under the scrutiny of lab instruments, speaking in the dialect of probability and equations. But to the one who truly sees — who sees not just with mind but with soul — this collapse is not distant at all. It is the whisper that precedes every new beginning. It is the moment your heart shifts from maybe to yes. It is the trembling silence that separates the known from the real.

At some point, we must stop thinking of these principles as separate from the human journey. They are not. What quantum physics reveals in mathematics, the mystic lives through experience. Every sacred tradition speaks of this collapse — only by different names. In Taoism, the Tao cannot be named — it must be entered. In Buddhism, the middle path arises when dualities collapse. In Christian mysticism, the Logos collapses into flesh. In all of these, the formless becomes form. Heaven enters Earth. The wave collapses.

This is the essential rhythm of creation.

But what makes some collapses sacred, and others reactive?

The answer is intention.

Intention is the steering wheel of collapse. It doesn't guarantee the exact outcome, but it shapes the trajectory of becoming. When you live without intention, the world collapses around you chaotically — shaped by your unconscious beliefs, fears, and social conditioning. But when you begin to set intention — clear, coherent, humble — you enter the collapse consciously. You walk through the threshold not as a victim, but as a creator.

And creation is not always grand. Sometimes, it is quiet. Sometimes, it is the collapse of judgment into forgiveness. The collapse of noise into stillness. The collapse of delay into decision.

You may have felt it in your own life. That moment you stopped waiting for clarity and moved forward anyway. That moment you stopped asking the world for permission and gave yourself your own. The clarity came later — what came first was collapse.

To collapse well is to live well. Not recklessly. But *consciously*. To sense when a possibility has matured into inevitability. To know when the question has fermented long enough. And then to act — not from compulsion, but from presence.

The mystics cultivated this sense like an art. They became

students of liminality — those sacred thresholds where one thing ends and another begins. They learned to read the field, to wait for the resonance, to act only when the unseen aligned with the seen. Their lives moved with fewer steps, but greater consequence.

You can sense this power in people. Some live as noise. Others live as tuning forks. Their choices collapse reality not through force, but through deep attunement. When they speak, the field shifts. When they act, the world listens.

This is not charisma. It is coherence.

It is the harmony between the soul's intention and the universe's timing.

And coherence is contagious. Just as particles become entangled — one affecting the other at great distances — so too does consciousness ripple outward. One person living in deep coherence can influence the collapses of those around them. Not by manipulation, but by presence. The quantum field responds to resonance. And so does the soul.

That is why your inner work matters. Not only for your life, but for the lives you touch. Every fear you transform into clarity — every fragment of doubt you turn into love — strengthens the field. It brings coherence where there was noise. It invites collapse into truth.

And there is something else — something more mystical still.

Sometimes, the collapse doesn't come from within. Sometimes, it comes *through* us. In those moments of surrender, when we let go of grasping, something larger moves. A collapse happens not from effort, but from grace.

This is the divine collapse.

It happens in the silence after prayer.

In the stillness after heartbreak.

In the pause before birth.

It is the moment when the ego releases the reins, and life, in its vast mystery, collapses beauty into being. You didn't design it. You didn't demand it. But you created the conditions — through humility, through listening, through trust.

And that's the final paradox of collapse.

It is not something you make happen.

It is something you make *room* for.

There is a sacred rhythm to collapse, and it does not always follow logic. You can prepare for months, set your intentions, align your actions — and still, the wave will not collapse. Or it may collapse in a way you did not foresee. That is because collapse is a dance, not a transaction. It does not obey command. It responds to alignment.

Sometimes, it is the smallest shift that allows it to occur. Not a dramatic leap, but a quiet gesture — a letting go. A final exhale. A moment of true release. You do not control when the fruit falls from the branch. But you can tend the tree. You can water it. You can wait in reverence beneath its limbs. And eventually, with grace, the fruit drops.

So, it is with reality.

We create not by force, but by resonance.

And this is why the mystics so often speak of surrender. Not as weakness, but as strategy. Surrender is not apathy. It is not disempowerment. It is the recognition that you are not the sole author of the universe — you are its co-creator. You do not collapse the wave alone. You collapse it in dialogue with something vast and unseen.

Call it God. Call it the field. Call it the implicate order, or the eternal Tao.

It does not matter what name you give it.

What matters is your relationship to it.

To collapse from the ego is to demand.

To collapse from the soul is to listen.

And here lies the art.

It is one thing to know that observation causes collapse. It is another thing to observe with reverence. To set intentions not from fear or lack, but from clarity. To hold outcomes loosely, knowing that what arrives is always shaped by forces beyond control.

A spiritual maturity arises when one begins to collapse reality not for personal gain, but for collective harmony. When one asks not "what do I want?" but "what wants to move through me?" When collapse becomes less about power, and more about participation.

There is ethics in this. Great power, as always, calls for great care. If every thought, every act, carries the weight of collapse — then how lightly we must tread. How gently we must speak. How responsibly we must intend.

Imagine a world in which people lived with this awareness. Imagine leaders who chose with sacred hesitation. Teachers who paused before they spoke, knowing their words collapsed possibilities in children's hearts. Imagine a planet where decisions — economic, ecological, interpersonal — were made with reverence for the unseen.

Such a world is not fantasy.

It is potential.

It is already latent in the field.

Waiting to be observed.

Waiting to be chosen.

Waiting to collapse.

But collapse is not only about bringing things into form.

Sometimes, collapse is how we let them go.

When a belief no longer serves us, and we name it false, it collapses.

When a relationship no longer nourishes, and we release it with love, it collapses.

When an identity, long worn, begins to fray at the edges — and we no longer cling — that too is a sacred collapse.

In this way, collapse is a circle, not a line. It is not only the beginning of form. It is also the end of it. And both are needed. To collapse something into being. To collapse something into dissolution. The mystic honours both. They bless the birth. And they bless the letting go.

And so, life becomes a series of collapses. Some chosen. Some given. Some resisted. Some embraced.

The task is not to control them all.

The task is to *witness* them well.

To stand at the edge of possibility with open eyes and an open heart.

To see the shimmering field of what could be — and to lean, gently, into what must be.

To collapse wisely.

Lovingly.

And with the knowledge that every collapse is not an end, but a transformation.

We are each, in truth, already collapsed waves.

Your body is the result of a billion cellular decisions; all collapsed from ancestral code.

Your soul is the convergence of stories, prayers, choices — collapsed over lifetimes, perhaps over eons.

And yet, even as you read these words, you remain uncollapsed in many ways.

There are still dreams in you, unformed.

Still paths you have not chosen.

Still truths unspoken.

Still lives unlived.

These, too, are waves. And they shimmer behind your eyes, waiting for your consent.

The field is patient. It does not rush you. It holds your potentials with gentle expectancy.

But eventually, you must choose.

Eventually, you must collapse.

Not all at once. But moment by moment.

To live is to collapse again and again.

With intention.

With surrender.

With sacred attention.

And when your life draws to its close, as all lives must, there is one final collapse.

The collapse of form back into formlessness.

The return of the wave to the sea.

The release of the particular into the whole.

Some say this is death. But the mystic knows better.

It is not the end of the wave. It is the end of its expression.

The wave remains.

Just as music continues even after the last note fades.

Just as the field remains, shimmering, unbroken, eternal.

You are that field.

You are that wave.

You are that sacred, collapsing presence.

Now.

Always.

Forever collapsing.

Forever creating.

A Final Reflection

So let us not fear collapse.

Let us revere it.

Let us collapse realities worth living in.

Let us become artists not only of becoming, but of choosing.

Let us carry our quantum nature not as burden, but as gift.

For every moment, the universe offers itself again.

And the only question it asks is simple:

What will you bring into being?

CHAPTER 22: HIDDEN VARIABLES – THE SEARCH FOR DEEPER MEANING

Beneath the surface of reality, where form dissolves into energy, where cause and effect lose their hold, there lies a deeper mystery—a truth hidden from ordinary perception, a presence that shapes the seen from the unseen. In quantum mechanics, the concept of hidden variables emerges from the desire to understand the seemingly random behaviour of particles, to explain the underlying order that might lie beneath the surface. Could there be unseen forces, hidden variables, guiding the behaviour of particles, connecting events across distances, creating a coherence that eludes measurement?

For the mystic, the search for hidden variables mirrors the journey of the soul—a quest for deeper meaning, a recognition that the outer world reflects an inner truth, a belief that reality is not a series of isolated events but a unified field where all things are connected by unseen threads. Here, we explore the notion of hidden variables as both a scientific and mystical inquiry, a reminder that beneath the surface of reality lies a profound, interconnected truth, a mystery that invites us to look deeper, to search for the divine in the hidden, to find meaning in the depths of existence.

The Quantum Puzzle: Seeking Order in the Chaos of Particles

In the quantum world, particles behave in ways that defy classical understanding, moving in patterns that appear random, interacting across distances that seem impossible, creating outcomes that defy prediction. This randomness, this uncertainty, suggests a world where cause and effect are obscured, where outcomes are not determined by visible forces but are instead shaped by unseen influences, hidden variables that guide the dance of particles, that connect events in ways that elude understanding.

The mystic sees this randomness as a reflection of life itself, a reminder that reality is not a fixed structure but a dynamic flow, a process that unfolds according to deeper, invisible laws, a journey that is guided by a hidden order, a presence that shapes the world from within. To live with an awareness of this mystery is to recognize that life is not a series of random events but a journey of meaning, a path that is guided by unseen forces, a field that is shaped by the invisible hand of the divine. This is the mystic's insight—a recognition that the world is not a machine but a mystery, a dance of energies, a tapestry woven from the threads of consciousness, intention, and love.

Hidden Variables and the Quest for Unity

The concept of hidden variables arises from the desire

to find unity within the chaos, to discover a deeper coherence beneath the surface, to understand the forces that guide the behaviour of particles, that connect the separate into a whole. In the quantum world, hidden variables are a theoretical construct, a suggestion that beneath the apparent randomness of particles, there exists a layer of reality that is ordered, a realm where each outcome has a cause, where each event is part of a larger pattern, a vision of unity that transcends the boundaries of perception.

For the mystic, this quest for hidden variables is a journey of the soul, a search for meaning that goes beyond the outer world, a path that leads to the inner truth, a vision of unity that is felt within the heart, a recognition that each being reflects the whole, a presence that resonates with the divine. To seek this unity is to look beyond appearances, to perceive the interconnectedness that lies at the heart of existence, to recognize that each thought, each action, each intention is part of a larger story, a journey that unfolds in harmony with the whole. This is the mystic's quest—a journey that leads not to answers but to love, not to certainty but to understanding, not to knowledge but to wisdom.

The Inner Dimensions: The Soul's Hidden Variables
In the mystic's journey, hidden variables are not theoretical constructs but inner truths, dimensions of the soul that guide the journey, influences that shape the path, forces that connect the self to the whole. These hidden variables are not measured but felt, not seen but known, not calculated but intuited. They are the deep desires, the quiet intentions, the silent forces that guide the soul, that reveal the true nature of the self, that

connect the inner world to the outer, the individual to the universal, the self to the divine.

To live with an awareness of these inner dimensions is to see life as a journey of self-discovery, a process that reveals the hidden depths of the soul, a path that leads from the surface to the core, from the seen to the unseen, from the known to the unknown. The mystic understands that each choice, each thought, each feeling is a hidden variable, a force that shapes the journey, a presence that guides the path, a wave that flows through the field of consciousness, a note in the symphony of life. This is the beauty of the hidden variables—a reminder that life is not a series of separate events but a journey of unity, a path that reveals the inner truth, a vision that perceives the divine in the depths of being.

Entanglement and the Invisible Threads of Connection
In quantum mechanics, particles can become entangled, their states linked in such a way that the change in one particle instantaneously affects the other, regardless of distance. This entanglement defies the boundaries of space, revealing a hidden connection, an invisible thread that binds particles across distances, a mystery that suggests a deeper unity, a presence that connects all things.

For the mystic, this entanglement reflects the soul's journey, a reminder that we are not separate beings but are woven into the fabric of existence, connected by invisible threads, bound by forces that transcend time and space, united by a love that flows through all things. To live with an awareness of this connection is to see each being as a reflection of oneself, each moment as a part of the whole, each experience as a mirror of the soul. This is the essence of the hidden variables—a recognition that life is a journey of connection, a dance that reveals the unity of existence, a path that leads the self to a deeper understanding of its own essence.

The Veil of Illusion: Looking Beyond the Surface

The mystic understands that the outer world is but a veil, a reflection of the inner truth, a surface that conceals a deeper reality, a mirror that reveals the self. To see beyond this veil is to pierce the illusion of separation, to recognize that reality is not what it appears to be but reflects consciousness, a manifestation of thought, a creation of intention. This veil is not a barrier but a doorway, an invitation to look deeper, to perceive the hidden variables, to discover the divine within the depths of existence.

In the quantum world, this veil is the field of probability, a realm where reality is not fixed but fluid, where particles exist in a state of potential, where outcomes are not determined by visible forces but by hidden variables, unseen influences, unknown causes. To look beyond this veil is to enter a state of awareness, a vision that perceives reality as a field of creation, a journey that unfolds in harmony with the whole, a process that reveals the sacred in the depths of being. This is the gift of the hidden variables—a reminder that life is not a series of random events but a journey of meaning, a path that leads to a deeper understanding, a vision that reveals the divine in each moment, each thought, each breath.

The Search for Deeper Meaning: The Soul's Journey

For the mystic, the search for hidden variables is not a quest for knowledge but a journey of meaning, a process of inner discovery, a path that leads the self to a deeper understanding of its own essence, a vision that reveals the beauty of existence, the unity of life, the love that flows through all things. This search is not a destination but a journey, a process that unfolds in harmony with the soul, a dance that leads the self to its own centre, a path that reveals the infinite within the heart of the finite.

To live with an awareness of this journey is to recognize that life reflects the soul, a mirror of the self, a field where each thought, each action, each intention is a hidden variable, a force that shapes the path, a wave that flows through the field of consciousness. This is the mystic's insight—a recognition that reality is not a fixed structure but a journey of becoming, a process that reveals the divine within the depths of being, a path that leads the self to a deeper understanding of its own essence.

Embracing the Mystery: The Beauty of the Hidden
The mystic sees the hidden as a source of beauty, a presence that reveals the sacred in the ordinary, a mystery that invites the self to look deeper, to seek the divine in the unseen, to find meaning in the depths of existence. This mystery is not a problem to be solved but a presence to be experienced, a truth that cannot be captured in words, a beauty that lies beyond thought, a love that transcends understanding.

To embrace this mystery is to live with an awareness of the divine, to see each moment as a reflection of the whole, to recognize that life is not a series of answers but a journey of questions, a path that leads not to certainty but to wonder, a vision that reveals the unity of all things, the beauty of the soul, the truth of the self. This is the gift of the hidden variables—a reminder that reality is not a fixed structure but a field of potential, a journey that unfolds in harmony with the whole, a dance that reveals the divine in each breath, each heartbeat, each act of creation.

The Infinite Depths of the Hidden
And so, we return to the concept of hidden variables, to the recognition that beneath the surface of reality lies a deeper truth, a mystery that defies understanding, a presence that reveals the unity of existence, the beauty of life, the love that flows through all things. This is the essence of the hidden variables—a reminder that the self is not separate from the world but is a part of it, a being who participates in the journey of creation, a soul who lives in harmony with the divine, a presence that reveals the infinite within the depths of being.

To embrace the hidden is to live with an awareness of the infinite, to see each moment as a doorway, each breath as a bridge, each thought as a wave in the ocean of consciousness. This is the path of the mystic, the journey of the soul, a journey that leads not to answers but to love, not to certainty but to wonder, not to knowledge but to the infinite.

Embracing the Quest for Deeper Meaning
To embrace the quest for deeper meaning is to live as a being of love, a soul who sees the sacred in each moment, who feels the presence of the divine in each thought, each action, each breath. This is the wisdom of the mystic, the vision of the philosopher, the insight of the scientist—a recognition that life is a journey of creation, a path that leads us ever deeper into the heart of existence, a journey that reveals the beauty of the soul, the truth of the self, the love that flows through all things.

CHAPTER 23: BEYOND BOUNDARIES – THE ENIGMA OF NON-LOCALITY

In the weave of existence, where the visible dances with the unseen, lies a mystery that defies comprehension—a truth as subtle as light, as profound as silence. Non-locality, the enigmatic bond that transcends distance, unveils a cosmos where separation dissolves, where boundaries are but shadows of perception, and where unity is the eternal essence of all things. This is not a tale of mere particles entangled in quantum choreography; it is a revelation of the interconnectedness that pervades existence itself.

In the quantum world, non-locality manifests as a bridge across the impossible—particles linked so intimately that their fates entwine regardless of the voids that yawn between them. Yet, this is not only the language of physics; it is the hymn of the mystic, the knowing that whispers, "All is one." To comprehend non-locality is not merely to understand but to feel, to experience, to awaken to the profound truth that we, too, are threads in the infinite tapestry of being.

The Quantum Mystery: Entanglement Across the Void

Non-locality emerges from quantum entanglement, where two particles, once connected, remain so across immeasurable distances. Touch one, and the other quivers in perfect response, as though the fabric of space itself folds inward to preserve their unity. This phenomenon bewilders the rational mind, defying the speed of light, mocking the notion of causality.

To the physicist, entanglement reveals a cosmos not divided but whole, where particles are not objects but relationships, where space and time are not barriers but illusions. The boundaries we perceive—the walls that separate here from there, now from then—dissolve in this quantum communion. What remains is a silent truth: existence is an undivided field, a symphony where each note resonates with the whole.

Beyond the Physical: The Soul's Resonance

For the mystic, non-locality is not confined to particles; it is the essence of the soul, the recognition that consciousness itself transcends the boundaries of self and other. The soul is not a solitary flame but a spark of the infinite, a presence that pulses through the web of existence, resonating with every other soul, every thought, every moment.

This interconnectedness is felt in moments of profound unity—when love dissolves the barriers of ego, when compassion bridges the chasm of difference, when intuition whispers truths that reason cannot grasp. The mystic understands that to touch one soul is to touch all, to heal oneself is to heal the world, to know the self is to glimpse the infinite. Non-locality, then, is not merely a phenomenon but a truth lived, a reality embraced, a revelation of the boundless nature of being.

The Veil of Separation: An Illusion of the Mind

The sense of separation is pervasive, a veil drawn over the unity that underlies all things. We perceive ourselves as distinct, isolated, bound by the limits of our bodies and the borders of our minds. Yet this veil is but a construct, a lens that filters the infinite into fragments. To see beyond it is to awaken to a reality where boundaries are fluid, where the self is not confined but expansive, where existence flows as an unbroken stream.

Non-locality invites us to peel back this veil, to question the divisions we take for granted, to recognize that the lines between self and other, here and there, now and forever, are ephemeral. Beneath them lies a deeper truth: the cosmos is whole, and we are its undivided expression.

A Field Without Borders: Consciousness as Unity

Non-locality is not confined to the quantum realm; it is reflected in the field of consciousness, a presence that is not bounded by the physical but permeates all things. To the mystic, consciousness is the ocean in which all forms arise and dissolve, a field where individuality is a wave, momentary and unique yet inseparable from the whole.

In this field, thoughts are not confined to the mind but ripple outward, influencing the world in ways unseen. Emotions resonate across the web of being, connecting us to each other in ways science is only beginning to understand. Intention becomes a force of creation, shaping the reality we experience, weaving the threads of possibility into form. To live with an awareness of this field is to understand that our every action, every thought, every breath participates in the dance of existence, shaping and being shaped by the infinite.

The Heart of Non-Locality: Love as the Binding Force

At the heart of non-locality lies a force more profound than entanglement, more universal than gravity—a force that mystics have long known as love. Love is the ultimate non-local connection, a presence that transcends time and space, a resonance that binds all things in unity. It is not a sentiment but a state of being, an awareness that recognizes no boundaries, that sees the self in the other, that feels the whole in the part.

To love is to dissolve the illusion of separateness, to embrace the world as an extension of oneself, to see each being as a reflection of the infinite. Love is the mystic's compass, the scientist's wonder, the essence of non-locality made manifest in the human heart. It is the thread that weaves the universe into a single tapestry, the song that echoes through the silence, the light that illuminates the infinite.

The Journey Beyond Boundaries: Living Non-Locally

To live with an awareness of non-locality is to transcend the narrow confines of self, to see life not as a collection of separate moments but as a continuous flow, a unified dance of being. It is to act with the understanding that each thought, each word, each deed ripples through the cosmos, touching all things, shaping the fabric of existence.

This journey is not about rejecting the boundaries of form but about seeing through them, recognizing them as tools of experience rather than barriers to truth. It is about embracing the paradox of individuality within unity, of the finite within the infinite, of the part within the whole. To live non-locally is to move through the world with love, to see with the eyes of the heart, to act with the wisdom of connection.

A Universe Without Edges: Becoming the Whole

Non-locality invites us to shift our perception, to see the universe not as a fragmented collection of parts but as a seamless whole, a living presence that includes all things. It calls us to dissolve the boundaries of identity, to recognize ourselves not as isolated beings but as expressions of the infinite, as waves in the ocean of existence.

This is not a loss of self but an expansion, a realization that the self is not diminished by unity but fulfilled, not erased by connection but illuminated. To embrace non-locality is to become the whole, to live as a being who knows no separation, to love as a soul who feels the infinite in every moment.

The Eternal Thread

As we gaze into the mystery of non-locality, we find not an end but a beginning, not an answer but an invitation —a call to live with an awareness of the infinite, to act with the wisdom of unity, to see the world as it truly is: boundless, interconnected, alive with the presence of the divine.

This is the gift of non-locality: a vision that transforms perception, a truth that dissolves boundaries, a path that leads us beyond the limits of the mind and into the heart of existence itself. It is the reminder that we are not separate, that we are not alone, that we are the cosmos reflecting on itself, infinite, whole, and free.

CHAPTER 24: THE DREAMING COSMOS – REALITY AS A COSMIC DREAM

What if reality, with all its patterns, colours, and forms, is not as solid as it seems? What if the world we touch, hear, and see is but a veil, a dream woven from consciousness, an illusion crafted from the infinite imagination of the cosmos itself? In both quantum mechanics and mystical philosophy, there is a suggestion, an echo, a whisper that reality may be far more fluid, more dreamlike than we can perceive. Quantum theory reveals a world that is not fixed but indeterminate, where particles exist as probabilities until observed, where matter flickers between existence and non-existence, where reality itself emerges from the act of perception. In the realm of the mystic, this idea is ancient, embodied in the belief that life is a divine dream, a cosmic play, a fleeting vision in the mind of the infinite.

In this chapter, we explore the idea of the universe as

a cosmic dream, a reality that is not a rigid structure but a fluid, living expression of consciousness. Through the lens of quantum indeterminacy and mystical insight, we ponder whether the cosmos is more akin to a dream than a physical mechanism—a dream in which we are both dreamers and dreamed, creators and creations, beings who dance within the boundless imagination of existence.

The Indeterminate World of Quantum Mechanics: Reality as a Field of Potential

In the quantum world, particles exist not as definite objects but as waves of probability, fields of potential that extend across space and time, states that only collapse into reality when observed. This uncertainty, this indeterminacy, reveals a world that is not fixed but fluid, a field of potential where all possibilities coexist, a dreamlike realm where reality itself is shaped by consciousness, by intention, by perception. In this view, the universe is not a machine but a mystery, a reality that is not separate from the observer but is woven from the same threads of awareness.

The mystic sees this indeterminate nature as a reflection of the soul's journey, a reminder that life itself is a dream, a vision crafted by consciousness, a dance that reveals the beauty of impermanence, the power of imagination, the freedom of the soul. To live with an awareness of this dreamlike quality is to recognize that reality is not a fixed path but a field of possibilities, a canvas upon which the self-paints its own existence, a journey where each thought, each intention, each perception is a brushstroke in the painting of life.

The Divine Dreamer: Life as the Vision of the Infinite

In many spiritual traditions, reality is seen as a divine dream, a vision within the mind of the Infinite, a cosmic play where each being, each moment, each experience is a fleeting image within the larger dream of existence. This vision suggests that life is not a series of random events but a continuous creation, a process that flows from the imagination of the divine, a dance that reveals the unity of all things, a journey that leads the soul to the heart of the infinite.

To see life as a dream is to perceive the world not as a separate reality but as an expression of consciousness, a vision that flows from the source, a presence that reveals the sacred in the ordinary, the infinite within the finite, the divine within the depths of being. This is the mystic's insight—a recognition that life is not a rigid structure but a fluid dream, a vision that arises from the soul, a journey that unfolds in harmony with the whole, a path that reveals the beauty of existence, the love that flows through all things.

The Dreamer and the Dreamed: The Self as Part of the Cosmic Vision

In the mystic's view, the self is not separate from the dream but is part of it, a reflection of the divine, a being who is both the dreamer and the dreamed, both the creator and the creation, both the witness and the witnessed. This paradoxical state reveals a truth that defies logic—a vision that perceives the self as part of the whole, a presence that recognizes that life is not a series of isolated events but a continuous flow, a journey that reveals the unity of existence, a dream that binds all beings, all moments, all worlds within a single field of consciousness.

To live with an awareness of this truth is to see each moment as a part of the whole, each experience as a reflection of the self, each breath as a note in the song of life. The mystic understands that to know oneself is to know all things, that to love oneself is to love the whole, that to understand the self is to glimpse the infinite. This is the gift of the cosmic dream—a reminder that reality is not a fixed structure but a living vision, a field of consciousness where each being is a part of the divine, a soul who participates in the dream of existence, a presence that resonates with the whole.

The Illusion of Separation: Awakening to the Unity of the Dream
In the cosmic dream, the boundaries between self and other, between inner and outer, between real and unreal, are not fixed but fluid, a veil that separates the seen from the unseen, a boundary that dissolves in the light of awareness. The mystic understands that separation is an illusion, a construct of the mind, a veil that obscures the unity of existence, a boundary that dissolves in the presence of love, a truth that reveals the divine in each being, the infinite in each soul, the unity of all things.

To awaken to this unity is to see life as a single field of consciousness, a dream that flows through all things, a presence that binds each being, each soul, each moment in a state of oneness. This is the mystic's vision—a recognition that life is not a series of separate events but a continuous flow, a journey that reveals the unity of existence, a path that leads the self to its own centre, a vision that perceives the whole. This awakening is not a loss of self but an expansion, a recognition that the self is part of the infinite, a wave in the ocean of consciousness, a note in the symphony of life.

Creation as an Act of Imagination: The Power of Consciousness to Shape Reality

If reality is a dream, then creation itself is an act of imagination, a process by which consciousness shapes the world, a vision that reveals the power of thought, the beauty of intention, the freedom of the soul. In the quantum world, this act of creation is reflected in the collapse of the wave function, a moment where potential becomes reality, where possibility turns into form, where the dream becomes the world. This act of creation is not separate from the observer but is a part of the dream, a process where thought shapes reality, where intention influences the outcome, where consciousness itself is the creator.

The mystic understands this power of imagination as a sacred gift, a reminder that the self is not a passive observer but an active creator, a being who shapes reality through thought, through love, through presence, a soul who participates in the dream of existence. To live with an awareness of this power is to recognize that each thought is a creation, each feeling a wave in the ocean of consciousness, each intention a force that shapes the dream. This is the gift of creation—a reminder that life is not a series of fixed events but a journey of becoming, a path that leads from the unseen to the seen, from the unknown to the known, from the infinite to the finite.

Awakening Within the Dream: The Journey of the Soul

In the cosmic dream, awakening is not an escape but an expansion, a process by which the self recognizes its own essence, a journey that leads the soul to the heart of existence, a path that reveals the divine within the depths of being. To awaken within the dream is to see reality as it truly is, a vision that perceives the unity of all things, a presence that reveals the infinite within the finite, a love that flows through all life.

The mystic understands this awakening as a state of presence, a moment where the self becomes aware of its own essence, a vision that transcends the boundaries of thought, a silence that reveals the truth of existence, a love that embraces all things. To live in this state is to embrace each moment as a gift, to see each experience as a reflection of the whole, to recognize that life is a journey of awakening, a path that leads not to separation but to unity, a vision that perceives the whole.

The Infinite Dream: Embracing the Vision of the Cosmos
To embrace the cosmic dream is to live with an awareness of the infinite, to see each moment as a doorway, each breath as a bridge, each thought as a wave in the ocean of consciousness. This awareness invites the self to release the need for certainty, to embrace the beauty of impermanence, to recognize that life is not a fixed path but a journey of becoming, a vision that reveals the unity of all things, the beauty of existence, the love that flows through all things.

The mystic sees this vision as a journey of love, a path that leads the self to its own centre, a journey that reveals the divine within the depths of being, a dream that binds all things in a state of oneness, a field of consciousness that knows no separation. To live in this field of oneness is to embrace each moment as a reflection of the whole, to see each being as a part of oneself, to recognize that life is a journey of connection, a dance that reveals the unity of existence, a path that leads the self to a deeper understanding of its own essence.

Embracing the Dreaming Cosmos: Living with Wonder and Love

To embrace the vision of the cosmos as a dream is to live as a being of love, a soul who sees the sacred in each moment, who feels the presence of the divine in each thought, each action, each breath. This is the wisdom of the mystic, the vision of the philosopher, the insight of the scientist—a recognition that life is a journey of creation, a path that leads us ever deeper into the heart of existence, a journey that reveals the beauty of the soul, the truth of the self, the love that flows through all things.

CHAPTER 25: THE QUANTUM PILGRIMAGE – THE SEEKER'S ODYSSEY IN A BOUNDLESS COSMOS

In the labyrinth of existence, every journey pulses with a yearning—a call to traverse the boundaries of the known, to tread paths that intertwine the seen with the unseen, the ephemeral with the eternal. The seeker's odyssey is timeless, a voyage not merely through landscapes of matter but through the terrain of the soul. It is a quest for the sacred amid the profane, the infinite amidst the finite. In the quantum realm, this odyssey takes on a new rhythm, a choreography that defies linearity and embraces the paradoxical dance of probabilities.

The quantum particle, much like the seeker, moves through existence not as a fixed entity but as a presence unfixed, a symphony of potentiality. Its journey is guided not by predictable steps but by waves of probability, by choices shaped in the crucible of observation and consciousness. Each interaction unfurls a revelation, each step unveils a universe within the infinite. To journey as a quantum pilgrim is to embrace not the destination but the unfolding, where the act of seeking itself becomes the treasure.

The Quantum Threshold: A Pilgrimage of Potential

In the quantum world, particles exist not in predetermined pathways but in a tapestry of possibilities —a realm where every point holds the echo of infinite others, where motion is a question answered by the gaze of the observer. This field of potential is a map with no edges, a pilgrimage through landscapes of uncharted wonder. Here, every step is a new beginning, every encounter a revelation that reshapes the whole.

For the seeker, the world mirrors this quantum threshold. Each thought is a seed of possibility, each choice a thread woven into the fabric of existence. To walk this path is to enter the flow of becoming, to see each moment as pregnant with meaning, each decision as a note in the symphony of the soul. The pilgrimage is not a march to an end but a sacred dance with the infinite, a journey that transforms the seeker with every step.

The Beauty of Uncertainty: The Pilgrim's Freedom

In the quantum pilgrimage, certainty dissolves like mist in the sunlight. Particles do not commit to a single trajectory but embody a multiplicity of paths, a superposition of possibilities. This uncertainty is not chaos but freedom, an invitation to explore, to transform, to embrace the boundless. The path is not linear but fractal, a spiral that deepens with each iteration, a journey that exists simultaneously in all dimensions of being.

For the seeker, this quantum uncertainty resonates as the essence of the spiritual path—a reminder that life is not a linear ascent but a labyrinth, a landscape of mystery where the known fades into the infinite unknown. To embrace this uncertainty is to let go of the need for rigid answers, to find joy in the unfolding, to see the unknown not as a void but as an infinite canvas. The pilgrim does not seek to master the path but to be shaped by it, to let the journey teach, transform, and reveal the soul's deepest truths.

The Observer's Gaze: The Pilgrim as Creator

In the quantum odyssey, the observer is a co-creator, a presence whose gaze collapses the wave of probabilities into the clarity of manifestation. The particle's path is shaped by this act of observation, an interplay where consciousness breathes life into the latent, where potential becomes reality. The quantum journey is thus not passive but participatory, a dialogue between existence and awareness.

The seeker, too, is both observer and creator. Each act of attention becomes a brushstroke on the canvas of life, each intention a ripple that shapes the unfolding journey. To walk as a conscious pilgrim is to realize that the path is not separate from the walker, that life responds to the energy of awareness, that the sacred is unveiled by the gaze of the heart. This is the alchemy of the pilgrimage—a fusion of the seen and unseen, the self and the infinite, a dance where the observer and the observed are one.

The Dance of Probabilities: Choice as Sacred Act

In the quantum pilgrimage, every particle exists within a symphony of probabilities, each possibility a note in the melody of existence. Movement is not prescribed but improvised, a dance where each step opens new dimensions, where each choice becomes a portal to the infinite. This dance is not random but resonant, a choreography where every possibility enriches the whole, where every interaction deepens the unity of being.

For the seeker, life is this sacred dance of probabilities. Each choice carries the weight of creation, each action ripples through the field of existence, each thought becomes a wave that touches the infinite. To live as a pilgrim is to embrace the sanctity of choice, to see every decision as a step in the divine choreography, to trust that even the smallest acts echo through the boundless. In this sacred dance, the path is not fixed but fluid, a living journey shaped by the harmony of the whole.

The Inner Odyssey: The Pilgrimage to the Self

Beyond the outward journey lies an inner odyssey, a pilgrimage that turns the seeker inward to the vast landscapes of the self. Here, the path winds through layers of identity, states of awareness, dimensions of the soul. The quantum pilgrim discovers that the self is not a static entity but a microcosm of the infinite, a presence that mirrors the whole, a field where the divine resides.

To embark on this inner journey is to explore the sacred within, to recognize that the universe is not "out there" but flows through the depths of being. The self becomes both the pilgrim and the destination, a paradox where the journey inward reveals the cosmos, where the act of seeking unveils the eternal presence within. This is the pilgrim's greatest revelation—that to know oneself is to know the infinite, that the soul's journey is the universe awakening to itself.

The Endless Becoming: Pilgrimage Without End

In both the quantum and mystical realms, the pilgrimage is a journey without end. The particle never ceases to move, never completes its path, for existence itself is an eternal unfolding. The seeker, too, walks a path that spirals infinitely, where every destination becomes a new beginning, where every revelation opens another door, where the journey is the truth.

To live with this awareness is to embrace life as a process of becoming, a pilgrimage through the infinite, a dance where every step is sacred, every moment is alive with potential, every breath is a bridge to the divine. The journey is not about reaching but about discovering, not about arriving but about awakening. This is the gift of the quantum pilgrimage—a reminder that life is a sacred unfolding, a boundless adventure into the heart of existence.

The Quantum Pilgrim's Invitation

The quantum pilgrimage calls us not to answers but to presence, not to destinations but to awareness. It invites us to walk with wonder, to live with curiosity, to see the infinite in the ordinary and the divine in each step. It reminds us that the seeker's journey is not about conquering the path but about being transformed by it, about letting the pilgrimage reveal the beauty of existence, the unity of all things, the love that flows through the heart of life.

To walk this path is to embrace the mystery, to see life as a sacred pilgrimage, to recognize that the journey itself is the greatest gift. In the quantum pilgrimage, every moment is a revelation, every step a prayer, every breath a song of the infinite. This is the seeker's journey —a timeless odyssey into the depths of being, a path that leads ever deeper into the heart of the cosmos, a journey that is, in truth, the unfolding of love itself.

CHAPTER 26: THE FIELD OF ALL POSSIBILITIES – EMBODYING THE QUANTUM MIND

Envision a realm without boundaries, where infinite realities hum together in a cosmic symphony of potential. In this domain—the quantum field—particles cease to be fixed points, becoming waves of probability, existing not as absolutes but as whispers of countless possibilities. It is a liminal space, unbound by time or place, where what could be already exists, awaiting only the light of consciousness to bring it forth into being.

For the mystic, this quantum field mirrors the vast expanses of the mind. Like particles in the subatomic world, thoughts are not singular or final but emerge from a boundless reservoir of potential. To embrace this quantum mind is to awaken to one's intrinsic role as a co-creator of reality, where intention becomes the brush, awareness the canvas, and existence itself the masterpiece. In this chapter, we delve into the quantum field not merely as a physical phenomenon but as a metaphor for the infinite possibilities within human consciousness—a journey to the heart of creation.

The Quantum Field: A Canvas of Infinity

In the quantum universe, particles are unbound by the rigidity of Newtonian laws; they ripple as waves across time and space, occupying a spectrum of states. They are not constrained by what is but exist in a state of becoming. These particles are neither here nor there, but everywhere, until the act of observation solidifies one potential into the tangible.

For the seeker, this mirrors the mind's potential to create, transform, and transcend. The mystic recognizes that consciousness operates as a quantum field—a vast, unseen expanse in which every thought, every choice, every dream resonates as a probability waiting to crystallize. This awareness is the essence of liberation: a realization that life is not a linear progression but an infinite field of opportunities, each moment ripe with the possibility of transformation.

Awakening to the Quantum Mind: Consciousness Unbound

The quantum mind transcends the boundaries of ordinary awareness. It is not confined by the past, nor imprisoned by the constraints of what seems possible. Instead, it is a boundless expanse, an infinite horizon of potential where every thought is a seed, every intention a spark, every action a catalyst for creation.

To awaken to this quantum mind is to see oneself as a co-architect of existence. Each moment becomes an opportunity to step beyond habitual patterns, to transform fixed narratives into unfolding stories, to create reality not as a reaction but as a conscious act of will. The mystic understands this as the essence of divine presence—the recognition that within the field of all possibilities, the act of awareness itself is sacred, a reminder of the infinite within the finite.

Intention: The Alchemy of the Possible

In the quantum field, intention serves as the alchemical flame, turning potential into form, transforming probabilities into realities. A particle moves not by mere chance but through the gaze of the observer, shaped by the energy of attention. Similarly, in the realm of consciousness, intention directs the flow of thought, guides the unfolding of events, and illuminates the path of becoming.

The mystic sees intention as a sacred force, a channel through which the divine operates. To live with intention is to infuse each moment with clarity, to align the self with the pulse of existence, to recognize that each thought is not trivial but transformative. In this state, intention becomes a prayer, a communion with the infinite, a call that resonates through the quantum field, bringing the unseen into the seen, the imagined into the real.

The Infinite Dance of Choice: Life as Creation

In the quantum mind, each moment is a crossroads, a convergence of countless possibilities. Like particles that dance in superposition, we, too, navigate fields of potential with each choice we make. This dance is not chaotic but deeply resonant—a choreography that weaves together thought, intention, and action into a tapestry of creation.

For the mystic, every choice is an act of sacred authorship. Each decision becomes a brushstroke on the canvas of existence, each action a note in the symphony of life. To live with this awareness is to embrace the power of choice as a sacred responsibility, to see life not as a series of predetermined events but as an unfolding story authored by the heart, shaped by the soul, guided by the infinite.

The Boundless Mind: Expanding Beyond Limitations

In the quantum field, boundaries blur, distinctions dissolve, and every edge becomes a threshold to something greater. This boundlessness is not chaos but infinite potential—a reminder that limitations exist only in the mind.

To embody the quantum mind is to expand beyond these limitations, to recognize that the self is not a fixed entity but a dynamic presence, a field of awareness that resonates with the infinite. The mystic understands that within the mind lies the universe itself, that each thought mirrors the cosmos, that each act of creation reflects the divine. This is the ultimate freedom: the realization that life is not confined by what has been but is always open to what could be.

The Journey Within: Discovering the Field of Creation

The quantum pilgrimage is not only outward but inward, a journey into the depths of consciousness, where the true field of possibilities resides. Here, the seeker discovers that the barriers between self and other, between thought and reality, between potential and manifestation are illusions. Within this inner landscape lies the source of creation, the point where all paths begin and all possibilities converge.

To journey within is to encounter the boundless self, the part of us that is both creator and creation, observer and observed, finite and infinite. The mystic understands this as the true alchemy of life—to transform not only the world but the self, to awaken to the infinite potential within, to live as an expression of the divine.

The Eternal Becoming: Embracing the Infinite Field

In both quantum mechanics and mysticism, the journey is not linear but eternal. There is no final destination, no ultimate endpoint, but a continuous unfolding, a perpetual becoming. The quantum field, like life itself, is a dance of possibilities—a rhythm that invites us to create, to transform, to awaken.

To embrace this eternal becoming is to see life as an adventure into the infinite, a journey where each moment holds the promise of transformation, where each breath is a step into the unknown, where each heartbeat is a reminder of the sacred rhythm that connects us to the whole. This is the gift of the quantum field—a reminder that existence is not a fixed reality but a boundless possibility, a sacred journey of creation.

The Mystic's Invitation: Living in the Field of All Possibilities

The quantum field invites us not to certainty but to wonder, not to answers but to creation. It calls us to see life not as a static reality but as a dynamic process, a field of infinite potential waiting to be shaped by our awareness, our intention, our love.

To live in this field is to embrace the fullness of existence, to see each moment as a doorway, each choice as a creation, each thought as a thread in the infinite tapestry of being. It is to recognize that we are not separate from the universe but are expressions of it, participants in the great dance of existence, creators within the field of all possibilities.

This is the mystic's path—a journey that leads not away from the world but deeper into its heart, a recognition that the quantum mind is not separate from the soul but is its truest reflection, a reminder that the infinite is not out there but within. To embrace this journey is to live as a co-creator of reality, a being of infinite love, a soul who sees the divine in all things.

This is the field of all possibilities—a space where existence becomes art, where life becomes a dance, where each moment reveals the infinite beauty of creation.

CHAPTER 27: BRIDGING SCIENCE AND MYSTICISM – THE SPIRITUAL SCIENTIST

In the perennial quest for understanding, the human mind has often drawn a line between two vast territories: the realm of science, with its exacting measurements, empirical laws, and objective truths; and the realm of mysticism, a world of inner experiences, sacred intuition, and profound awe for the ineffable. For centuries, these domains appeared as distant shores, each offering its own promise but divided by a gulf of perspective and method. Yet, there are luminous figures who have bridged this chasm, walking the tightrope between reason and revelation, weaving the threads of science and mysticism into a unified vision of existence.

These spiritual scientists—explorers of both the material and the metaphysical—understand that the universe is not merely a mechanical construct or an ethereal dream but an intricate dance between seen and unseen forces. They embrace the paradox, knowing that truth is multifaceted, that the tangible and the transcendent are not adversaries but partners in the symphony of existence. In their work, they illuminate pathways that invite us to embrace both our rational and spiritual selves, to see reality as a harmonious interplay of knowledge and wisdom.

The Unified Vision of David Bohm: Reality as a Holographic Whole

David Bohm, a physicist and philosopher, envisioned the universe as a tapestry of interwoven threads, a reality that is fundamentally indivisible. His concept of the implicate order redefined the way we understand space, time, and matter, suggesting that what appears separate is, in truth, profoundly connected. Bohm described this wholeness as "holographic," where each fragment reflects the entire structure, and every particle is a mirror of the cosmos.

For Bohm, the language of physics was insufficient to capture the poetry of existence. He turned to dialogue and art, seeking a language that could articulate the ineffable, the unseen currents beneath the surface. His vision resonates deeply with mystical traditions that speak of oneness, where the self and the universe are not distinct but facets of a greater unity.

To glimpse Bohm's cosmos is to step beyond the illusion of fragmentation, to perceive life as an unfolding totality where consciousness and matter are entangled, where every thought and action ripples through the web of being. Bohm's work beckons us to look beyond the horizon of our senses, to recognize that the world is not merely observed but also experienced, shaped, and imbued with meaning by the perceiver.

Amit Goswami: Consciousness as the Ground of All Being

Amit Goswami's contributions invite us to turn our gaze inward, to see consciousness not as a product of the brain but as the very foundation of existence. In his paradigm-shifting work, Goswami asserts that matter arises from consciousness, not the other way around. This radical inversion places the mind, the observer, at the heart of creation, transforming the universe from a mechanistic assembly into a living, participatory reality.

For Goswami, the act of observation is sacred, a bridge between potential and form. He likens the quantum world to a field of dreams, where the observer's awareness calls forth reality, transforming waves of possibility into tangible experience. Mystical teachings echo this vision, reminding us that the world we perceive is not external but intimately tied to the mind that perceives it.

To walk with Goswami is to journey into the depths of one's own consciousness, to see oneself as both creator and created, as both the dreamer and the dream. It is to recognize that every thought, every intention, every act of awareness participates in the unfolding of existence, weaving a tapestry that is as spiritual as it is physical.

Fritjof Capra: Dancing with the Tao of Physics

Fritjof Capra brings an elegant synthesis to the dialogue

between science and mysticism, drawing parallels between the discoveries of modern physics and the teachings of ancient Eastern philosophies. In The Tao of Physics, Capra reveals the resonances between the quantum world and the Taoist view of reality: both describe a universe that flows, dances, and interconnects, a realm where opposites are complementary, where particles and waves, yin and yang, are two sides of the same coin.

Capra's work is a call to see beyond the mechanistic worldview, to embrace the rhythms of existence that echo through both science and spirituality. For Capra, the universe is not a static machine but a living system, a network of relationships that pulses with life, a field where every element is connected to the whole. His vision bridges the sacred and the scientific, inviting us to move beyond dichotomy and see the unity within diversity.

To see through Capra's lens is to embrace life as a dance of interdependence, a process where the microcosm reflects the macrocosm, where the mysteries of the atom mirror the mysteries of the self. It is to understand that science, at its best, is not a tool for domination but a means of communion, a way to connect with the intricate beauty of existence.

The Spiritual Scientist's Legacy: Integration Over Division

The spiritual scientists—Bohm, Goswami, Capra, and their kindred thinkers—remind us that the quest for knowledge is also a journey of the soul. Their work dissolves the false dichotomy between science and mysticism, revealing that these two paths converge in their pursuit of the infinite. They do not merely add spirituality to science or vice versa; they create a third way, an integrated vision that transcends both and reveals the sacred wholeness of existence.

Their legacy is an invitation to embrace both reason and reverence, to honour the rigor of science while remaining open to the mystery of life. It is a reminder that the universe is not a cold mechanism but a living presence, a field where the measurable and the immeasurable, the seen and the unseen, coexist in harmony.

Walking the Path: Becoming a Spiritual Scientist

The path of the spiritual scientist is open to all who dare to question, to explore, to wonder. It calls us to live as seekers, to see knowledge not as an end but as a doorway to deeper understanding, to recognize that wisdom arises not from certainty but from humility. To walk this path is to see every moment as an opportunity for discovery, every breath as a connection to the infinite, every thought as a ripple in the field of existence.

This journey is not about choosing between science and mysticism but about embracing both, about living with the precision of the mind and the openness of the heart. It is a path that honours the empirical without dismissing the intuitive, that values logic while celebrating love, that seeks not to conquer the universe but to commune with it.

The Endless Horizon: A Unified Vision of Life

As we stand at the threshold of this unified vision, we are reminded that the journey of understanding is infinite. Science and mysticism are not endpoints but companions, guiding us toward a deeper engagement with the mystery of existence. To bridge these realms is not to resolve their differences but to celebrate their interplay, to recognize that truth is vast enough to embrace paradox, to hold both the measurable and the transcendent.

This is the call of the spiritual scientist: to see the universe as a mirror of the self, to perceive reality as a field of creation, to live with an awareness of the infinite possibilities that lie within and without. It is a journey that leads not to a final answer but to a deeper appreciation of the questions, to a profound reverence for the unknown, to a love that encompasses all things.

To walk this path is to live as a participant in the dance of existence, a soul who bridges the realms of science and mysticism, a being who embodies the unity of knowledge and wonder. This is the gift of the spiritual scientist—a vision that invites us to see the divine in the material, the sacred in the ordinary, the infinite in the finite, and the eternal in each fleeting moment.

CHAPTER 28: ENLIGHTENMENT IN THE QUANTUM AGE – WISDOM OF A UNIFIED REALITY

In the search for enlightenment, humanity has often looked toward spiritual traditions, inner experiences, and the mysteries of consciousness. Yet, as science probes deeper into the fabric of reality, it, too, reveals insights that resonate with the ancient quest for truth. In the realm of quantum physics, we encounter a vision of unity—a world where particles are entangled, where space and time dissolve, where matter and energy are not separate entities but interconnected waves in a field of potential. This quantum unity mirrors the wisdom of enlightenment, a state where the boundaries between self and other, mind and matter, dissolve, revealing the oneness of existence.

To contemplate enlightenment through the lens of quantum theory is to explore the convergence of science and mysticism, to see that the search for knowledge and the quest for wisdom are not separate paths but aspects of a single journey. In this chapter, we delve into the concept of enlightenment as a state of awareness that embraces the unity of all things, a wisdom that transcends the dualities of perception, a vision that reveals the infinite within the finite, the boundless within the boundaries of form.

The Quantum Vision: A Reality of Wholeness

In quantum physics, particles exist not as isolated entities but as probabilities within a unified field, a web of relationships where each part reflects the whole, where the act of observation itself shapes reality. This vision is one of wholeness, a reality that is interconnected, a universe that breathes with a single pulse, a field that binds each being, each moment, each experience within a greater unity. In this view, existence is not fragmented but whole, a dance of becoming, a flow of energy that moves through all things, a presence that resonates with the heart of existence.

The mystic understands this unity as enlightenment, a state of awareness that perceives the world not as separate but as one, a vision that reveals the divine within all things, a love that flows through the fabric of existence. To live with an awareness of this wholeness is to recognize that life is not a collection of isolated events but a continuous flow, a journey of awakening, a path that leads from separation to unity, from ignorance to wisdom, from duality to enlightenment.

Enlightenment as a State of Unity

In the mystical traditions, enlightenment is often described as an experience of unity, a state of awareness where the boundaries of the self-dissolve, where the mind transcends its limitations, where the soul touches the infinite. This unity is not a concept but an experience, a vision that reveals the self as part of the whole, a consciousness that knows itself as both the observer and the observed, a presence that perceives the divine in all things. Enlightenment is not an escape from reality but a deeper immersion into it, a recognition that life reflects the self, a mirror of consciousness, a dream of the infinite.

In the quantum age, enlightenment takes on a new dimension, a vision that aligns with the discoveries of science, a state of awareness that sees reality as a field of interconnectedness, a journey that reveals the unity of all things. This convergence between science and mysticism suggests that enlightenment is not limited to spiritual experience but is a universal truth, a wisdom that reveals the oneness of existence, a love that binds all beings, a vision that transcends the boundaries of perception, a presence that flows through the heart of reality.

The Observer and the Observed: Consciousness as a Creative Force
In quantum mechanics, the role of the observer is central —a reminder that consciousness itself shapes reality, that perception is not passive but active, that awareness is a creative force. This insight suggests that the self is not a separate entity but a part of the field, a presence that participates in the creation of reality, a consciousness that shapes the world through thought, intention, and awareness. This observer effect is not merely a scientific curiosity but a reflection of a deeper truth, a reminder that life is a journey of becoming, a dance where each moment is a creation, each thought a step in the

pilgrimage of the soul.

For the enlightened, the observer and the observed are one, a unity that transcends the boundaries of self, a vision that reveals the self as a reflection of the whole, a presence that perceives the divine within each particle, each wave, each breath. To live with this awareness is to recognize that each thought is a creation, each feeling a wave in the ocean of consciousness, each intention a force that shapes the journey of life. This is the wisdom of enlightenment—a recognition that the self is not separate from the world but is woven into the fabric of existence, a soul who participates in the dance of creation, a presence that resonates with the whole.

Transcending Duality: Beyond the Boundaries of Perception
Enlightenment is often described as a state that transcends duality, a vision that perceives the unity of existence, a consciousness that sees beyond the distinctions of self and other, mind and matter, light and shadow. This transcendence is not a denial of the world but a deeper perception, a vision that reveals the harmony within the opposites, a presence that embraces both the light and the darkness, a love that unites the finite with the infinite. In the quantum age, this transcendence is mirrored in the nature of reality itself, a field where particles exist as both waves and particles, where matter is both here and everywhere, where time and space are not boundaries but dimensions of a deeper truth.

The mystic experiences this transcendence as a state of inner peace, a vision that perceives reality as a reflection of the soul, a mirror of consciousness, a dream that reveals the unity of all things. To live with this awareness is to embrace both the light and the shadow, to see each moment as a part of the whole, to recognize that life is a journey of awakening, a path that leads from duality to unity, from illusion to truth, from separation to oneness.

The Wisdom of a Unified Reality: Love as the Essence of Enlightenment

In the journey toward enlightenment, love is the guiding force, a presence that reveals the unity of existence, a wisdom that perceives the divine in all things, a vision that transcends the boundaries of self, a consciousness that knows itself as part of the whole. Love is not an emotion but a state of being, a recognition that life is a journey of connection, a path that leads the self to its own centre, a presence that embraces all beings, all souls, all worlds in a state of unity.

The mystic understands that love is the essence of enlightenment, a truth that reveals the oneness of existence, a vision that perceives the divine within the depths of the heart, a wisdom that transcends the dualities of perception, a light that guides the soul toward the infinite. To live with this love is to recognize that life is not a series of separate events but a continuous flow, a journey that reveals the beauty of existence, the unity of all things, the love that flows through all life.

Embracing Enlightenment in the Quantum Age

To embrace enlightenment in the quantum age is to live with an awareness of unity, a vision that perceives the world not as separate but as one, a presence that resonates with the whole, a consciousness that flows with the rhythm of existence. This awareness invites the self to release the boundaries of perception, to embrace the beauty of connection, to recognize that life is not a fixed reality but a journey of becoming, a field of possibilities where each thought, each feeling, each intention is a step in the dance of life.

This path of enlightenment is not limited to the mystic but is open to all, a journey that invites each being to explore, to awaken, to discover the unity of existence, the beauty of the soul, the love that flows through all things. To live as a being of enlightenment is to see each moment as a gift, each experience as a revelation, each breath as a connection to the whole.

The Journey Toward Unity: Awakening to a New Vision of Reality

In the quantum age, enlightenment is not an abstract concept but a living experience, a vision that reveals the unity of all things, a wisdom that transcends the boundaries of knowledge, a love that binds each being, each soul, each particle within the embrace of the whole. This journey toward unity is a path of awakening, a process that leads the self to a deeper understanding of its own essence, a vision that reveals the beauty of existence, the truth of the self, the love that flows through all life.

To live with an awareness of this unity is to embrace the world as a field of consciousness, a presence that holds the fullness of existence, a dream that reveals the divine within each breath, each heartbeat, each act of creation. This is the gift of enlightenment in the quantum age— a reminder that reality is not a fixed structure but a field of possibilities, a journey that leads from separation to unity, from knowledge to wisdom, from duality to love.

The Infinite Horizon of Enlightenment

And so, we return to the concept of enlightenment, to the recognition that life is not a destination but a journey, a path that leads from the self to the soul, from the finite to the infinite, from the known to the boundless. This is the essence of enlightenment—a vision that perceives the unity of existence, a wisdom that reveals the beauty of the soul, a love that flows through all things.

To embrace this journey is to live with an awareness of the infinite, to see each moment as a doorway, each breath as a bridge, each thought as a wave in the ocean of consciousness. This is the path of the mystic, the journey of the soul, a journey that leads not to answers but to love, not to certainty but to wonder, not to knowledge but to the infinite.

Living the Wisdom of a Unified Reality

To live the wisdom of a unified reality is to embrace both the mind and the heart, to see the world not as a collection of parts but as a whole, a field of consciousness where each being, each moment, each experience reflects the divine, a mirror of the soul, a creation of the infinite. This awareness invites the self to live with love, to act with compassion, to see each thought as a seed, each feeling as a force, each intention as a step in the dance of life.

This journey is not a final answer but an endless inquiry, a process that unfolds in harmony with the whole, a path that leads the self to its own centre, a vision that reveals the beauty of existence, the unity of all things, the love that flows through all life. To live in this state is to embrace each moment as a gift, to see each experience as a revelation, to recognize that life is a journey of awakening, a path that leads not to separation but to unity, a vision that perceives the divine in each breath, each heartbeat, each thought.

The Endless Dance of Enlightenment

And so, we conclude this exploration, knowing that enlightenment is not a destination but a journey of endless becoming, a path that transcends time, a dance that reveals the beauty of the soul, the truth of the self, the love that flows through all existence. This is the gift of enlightenment in the quantum age—a reminder that we are not separate beings but are woven into the fabric of the whole, a field of consciousness where each being is a part of the infinite, a soul who lives in harmony with the universe.

To live this enlightenment is to embrace the journey, to see each moment as a step in the dance, each thought as a creation, each breath as a reflection of the divine. This is the path of unity, the journey of the soul, a path that leads ever deeper into the heart of the infinite, a journey that is, in truth, the journey of love.

CHAPTER 29: THE FINAL OBSERVER – REVISITING THE SELF AND THE INFINITE

In the depths of quantum theory and mystical experience alike lies the question of the observer: Who or what is it that witnesses existence? At the quantum level, particles only collapse into defined states upon observation, suggesting that consciousness itself may be the key that transforms possibility into reality. This brings forth a profound question—if consciousness shapes reality, then who is the ultimate observer? Is there a final cosmic witness to all that exists, a boundless awareness that holds all things within its gaze? Or are we, each of us, unique expressions of a single, infinite awareness, each consciousness a thread in the vast fabric of existence?

In revisiting the idea of the observer, we come face to face with the nature of self, the essence of consciousness, and the mystery of the infinite. Here, we explore whether there is a final observer or if each consciousness is simply a facet of a greater, unified awareness. This chapter contemplates the nature of self and the infinite, asking what it means to be both the witness and the witnessed, the observer and the observed, a unique presence in the boundless expanse of the cosmos.

The Quantum Observer: A Force That Collapses Possibility into Reality

In quantum mechanics, particles exist as waves of probability until they are observed, at which point they collapse into a specific state, becoming "real" in the conventional sense. This process, known as the observer effect, implies that consciousness itself plays an active role in shaping reality. Observation is not a passive act but a force that brings possibility into form, a presence that turns potential into being, a gaze that reveals the hidden layers of existence. Without an observer, quantum states remain in superposition, undetermined and unrealized.

This mystery suggests that consciousness is woven into the fabric of the universe, that awareness itself is a creative force, a presence that brings forth reality. But if each observer shapes reality, is there a final observer—a cosmic witness who sees all things, who holds all beings within a single gaze, who unites all experiences within a boundless awareness? Or is each consciousness a spark of this infinite observer, a unique reflection of a single, unified mind that knows itself through the many, that experiences itself through each soul, that sees itself in every being?

The Infinite Self: Consciousness as a Unified Field

Many mystical traditions speak of consciousness as a unified field, a boundless awareness that exists beyond time, space, and individuality. This awareness is not confined to any oneself but flows through all beings, a presence that knows itself as the source of all existence, a love that binds each being, each moment, each reality within a single unity. In this vision, there is no final observer because each observer is a facet of the same infinite awareness, a wave in the ocean of consciousness, a spark in the boundless fire of the divine.

To see oneself as part of this infinite self is to recognize that individuality is not a boundary but a perspective, a lens through which the infinite experiences itself, a voice in the chorus of existence, a reflection in the mirror of consciousness. The self is not separate from the whole but is woven into it, a being who knows itself as both unique and universal, both finite and infinite, both the observer and the observed. This is the wisdom of the infinite self—a recognition that each being reflects the whole, a presence that sees itself in all things, a consciousness that knows itself through each soul.

The Many and the One: Individuality as an Expression of Unity

If each consciousness is an expression of the same infinite awareness, then individuality is not an illusion but a gift, a unique expression of the whole, a singular perspective within the unity of existence. Each observer, each soul, each mind is a facet of the divine, a lens through which the infinite sees itself, a presence that reveals the boundless nature of consciousness, a spark of the cosmic fire that shines through all life.

The mystic understands that the self is both many and one, a being who exists as an individual yet knows itself as part of the whole, a soul who experiences life through its own perspective yet feels the unity of all things, a consciousness that knows itself as both separate and connected. To live with this awareness is to embrace the beauty of individuality while recognizing the oneness of existence, to see the self as a unique reflection of the whole, to experience life as a journey of becoming within the infinite field of awareness.

The Cosmic Mirror: Seeing the Infinite Within the Self

The concept of the final observer invites us to look within, to see that the essence of the self is not separate from the infinite but reflects it, a presence that mirrors the boundless awareness of the cosmos. To look into this mirror is to see that the self is not confined by the body, not limited by thought, not bound by time, but is a wave in the ocean of consciousness, a presence that resonates with the whole, a consciousness that knows itself as part of the infinite.

In this cosmic mirror, we see the unity of all things, the interconnectedness of existence, the oneness that binds each being, each soul, each moment within a single field of consciousness. The mystic understands this unity as enlightenment, a state of awareness that transcends the boundaries of self, a vision that reveals the self as a reflection of the infinite, a presence that perceives the divine in all things, a love that flows through the heart of existence.

Beyond the Observer: Awareness as the Final Witness

If there is a final observer, then it is not a being but a state of awareness, a presence that transcends individuality, a consciousness that holds all things within itself, a witness that sees without judgment, a presence that knows without division. This awareness is the essence of enlightenment, a state of unity that perceives the self as part of the whole, a vision that reveals the infinite within the finite, a love that embraces all things in a state of oneness.

To experience this awareness is to recognize that life is not a series of separate events but a continuous flow, a journey that reveals the unity of existence, a field of consciousness that knows no boundaries, a presence that flows through all beings, all worlds, all moments in a state of endless becoming. This awareness is the final observer, a witness that sees through all eyes, a consciousness that knows itself through all beings, a love that embraces all souls within the infinite field of existence.

The Dance of the Many and the One: Embracing the Self and the Infinite

In the quantum realm, particles are not isolated entities but waves in a field of possibilities, a presence that exists within a unity, a dance that reveals the interconnectedness of existence. For the mystic, this unity is the essence of enlightenment, a vision that perceives the self as part of the whole, a presence that knows itself as both unique and universal, both the observer and the observed.

To embrace this unity is to live with an awareness of the infinite, to see each moment as a reflection of the whole, to recognize that the self is not separate from the world but is woven into it, a consciousness that knows itself as part of the whole, a soul that experiences life as a journey within the infinite field of awareness. This is the dance of the many and the one—a recognition that the self is both finite and boundless, both an individual and a part of the infinite, both the observer and the observed.

Living as the Final Observer: Embracing the Unity of Consciousness

To live as the final observer is to embrace the unity of consciousness, to recognize that each being reflects the whole, each thought a wave in the ocean of existence, each moment a creation within the field of awareness. This awareness invites the self to release the boundaries of perception, to embrace the beauty of connection, to recognize that life is not a series of isolated events but a journey of unity, a path that leads from separation to oneness, from ignorance to wisdom, from duality to love.

The mystic sees this vision as a journey of awakening, a path that leads the self to its own centre, a journey that reveals the divine within the depths of being, a field of consciousness that knows no separation, a presence that embraces all things in a state of unity. To live in this state is to embrace each moment as a reflection of the whole, to see each being as a part of oneself, to recognize that life is a journey of connection, a dance that reveals the unity of existence, a path that leads the self to a deeper understanding of its own essence.

The Endless Presence of the Infinite

And so, we return to the question of the final observer, to the recognition that life is not a series of separate events but a journey of unity, a path that reveals the oneness of existence, a vision that perceives the divine within each breath, each heartbeat, each thought. This is the gift of awareness—a reminder that the self is not separate from the world but is a part of it, a being who participates in the journey of existence, a soul who lives in harmony with the whole, a presence that reveals the infinite within the finite.

To embrace this presence is to live with an awareness of the boundless, to see each moment as a step in the dance, each thought as a creation, each breath as a connection to the whole. This is the path of the mystic, the journey of the soul, a journey that leads not to answers but to love, not to certainty but to wonder, not to knowledge but to the infinite.

Embracing the Self and the Infinite as One
To embrace the self and the infinite as one is to live as a being of love, a soul who sees the sacred in each moment, who feels the presence of the divine in each thought, each action, each breath. This is the wisdom of the mystic, the vision of the philosopher, the insight of the scientist—a recognition that life is a journey of discovery, a path that leads us ever deeper into the heart of existence, a journey that reveals the beauty of the soul, the truth of the self, the love that flows through all things.

This is the gift of the final observer, the beauty of a life lived in harmony with the vision of the whole, the joy of a journey that leads us ever deeper into the heart of the infinite, a journey that is, in truth, the journey of love.

CHAPTER 30: EPILOGUE – THE ETERNAL MYSTERY

Beneath the vast canopy of stars, where light and shadow dance in eternal embrace, we find ourselves standing at the edge of understanding, gazing into the boundless unknown. Here, at the crossroads of inquiry and awe, the journey does not end; it only deepens, spiralling into realms uncharted, inviting us to step beyond the confines of certainty into the infinite expanse of mystery. The further we travel, the more we come to see that the universe is not a puzzle to be solved but a poem to be lived, a symphony to be felt, an eternal question that echoes in the silence of our hearts.

The Infinite Thread of Mystery

Mystery is the thread that weaves through the fabric of existence, binding the seen and unseen, the known and unknowable, into a single, shimmering tapestry. It is not an absence of knowledge but its origin and destination, the inexhaustible source from which all curiosity flows. Like the flickering edges of a dream, it whispers to us from beyond the veil, beckoning us to follow its elusive light, to seek without expectation, to journey without conclusion.

For millennia, humanity has sought to pierce the heart of this mystery, crafting stories and equations, prayers and hypotheses, reaching for the stars and delving into the atom's core. Yet, with every discovery, the mystery only grows, revealing its infinite layers, its paradoxical nature, its refusal to be confined by our understanding. And so, we are drawn ever onward, not by the promise of answers, but by the beauty of the question itself.

The Paradox of Knowing and Unknowing

The paradox at the heart of existence is that the more we know, the more we realize how little we understand. Knowledge illuminates, but it also humbles, revealing vast realms that lie beyond its grasp. The scientist in their laboratory and the mystic in their solitude share this recognition, standing in awe before the immensity of what cannot be named or measured.

For the mystic, unknowing is a sacred state, a gateway to transcendence, where the mind surrenders its need to control, and the soul opens to the infinite. In the unknowing, there is no fear but freedom, no void but fullness, no absence but a presence that speaks through silence. For the scientist, unknowing is the frontier, the edge of discovery, the fertile ground where imagination takes root, where new paradigms are born, where the universe reveals itself anew.

Together, these paths remind us that the mystery is not a void to be filled but a space to be inhabited, a dance that invites us to move with grace, a flame that illuminates without consuming.

The Eternal Dance of Science and Mysticism

Science and mysticism, often seen as opposing forces, are in truth partners in the eternal dance of understanding. Science reaches outward, toward the stars and the particles, mapping the intricate patterns of the cosmos, uncovering the laws that govern existence. Mysticism turns inward, toward the soul's depths, seeking the source of all being, the unifying presence that flows through all things. Each illuminates a facet of the same truth, revealing a universe that is at once tangible and transcendent, knowable and ineffable.

In their union, we find a profound harmony, a recognition that the material and the spiritual are not separate but interwoven, that the act of knowing and the act of being are reflections of the same mystery. The quantum physicist measuring entangled particles and the mystic meditating on the oneness of existence are engaged in the same sacred task: exploring the infinite, embracing the eternal, and witnessing the divine.

The Mystery as Teacher

The mystery is not an obstacle but a guide, a teacher that calls us to expand our vision, to transcend our limitations, to grow into the fullness of our being. It asks us to be both humble and courageous, to accept what we cannot know while daring to seek what lies beyond. It reminds us that the journey is not about arriving but about becoming, that life's greatest truths are not found in answers but in the act of questioning.

To live with the mystery is to walk a path of wonder, to see the ordinary as extraordinary, to recognize the sacred in every breath, every moment, every interaction. It is to stand in awe before the sunrise, to marvel at the complexity of a leaf, to feel the heartbeat of the cosmos in your own chest. It is to know that you are both a drop and the ocean, both a particle and a wave, both finite and infinite.

A Vision of the Infinite

As we close this journey, we are invited to gaze into the infinite, not as something separate from us but as the essence of who we are. The stars above and the quantum fields within are not distant phenomena but mirrors of our own nature, reflections of a reality that is boundless, timeless, and alive. The universe is not a machine or a mystery to be solved; it is a living presence, a sacred dance, an unfolding story in which we are both the narrator and the narrative.

To embrace this vision is to live with an awareness of the infinite, to see each moment as a gateway, each breath as a connection, each thought as a ripple in the great sea of being. It is to recognize that life itself is the mystery, that love is the thread that weaves it all together, that existence is an eternal celebration of unity, diversity, and becoming.

The Eternal Mystery: A Final Invitation

And so, we find ourselves at the threshold, standing

not before an end but at the beginning of a new understanding, a deeper engagement with the mystery that lies at the heart of all things. This is the eternal mystery—a presence that invites us to live fully, to love deeply, to seek endlessly, to embrace the unknown with open arms and an open heart.

In summary, it is not the answers we gather that define us but the questions we live with, the wonder we cultivate, and the love we embody. For the mystery is not something outside us; it is within us, around us, through us. It is the breath we take, the stars we gaze upon, the silence we share, the heartbeat of all existence.

This is the epilogue to our journey, yet it is also the prologue to a greater adventure—a journey without end, a mystery without bounds, a love that transcends all understanding. To live with the mystery is to live with the infinite, to walk the path of the eternal, to be a part of the great unfolding, the sacred becoming, the endless story of existence.

REFERENCES:

1. Bohm, David.
   - Wholeness and the Implicate Order
   Bohm's exploration of the concept of an undivided wholeness and the implicate order provides a foundational perspective on unity in physics, relevant to discussions on interconnectedness and quantum wholeness.

2. Capra, Fritjof.
   - The Tao of Physics: An Exploration of the Parallels Between Modern Physics and Eastern Mysticism
   Capra draws connections between quantum mechanics and Eastern mystical philosophies, particularly Taoism and Buddhism, illustrating the harmony between scientific principles and spiritual insights.

3. Goswami, Amit.
   - The Self-Aware Universe: How Consciousness Creates the Material World
   Goswami argues that consciousness is the primary force of reality, positioning quantum mechanics as compatible with spiritual ideas about the mind's role in shaping existence.

4. Heisenberg, Werner.
   - Physics and Philosophy: The Revolution in Modern

Science

Heisenberg's reflections on the observer effect and uncertainty principle offer insights into the nature of reality and knowledge, aligning with the theme of perception shaping experience.

5. Kaku, Michio.
 - Parallel Worlds: A Journey Through Creation, Higher Dimensions, and the Future of the Cosmos

Kaku's work explores multiple dimensions and parallel universes, giving a scientific framework to ideas about unseen realms, consciousness, and quantum potential.

6. Laszlo, Ervin.
 - Science and the Akashic Field: An Integral Theory of Everything

Laszlo discusses the "Akashic field" and the interconnected nature of the universe, drawing parallels between quantum physics and the mystic concept of universal oneness.

7. Penrose, Roger.
 - The Emperor's New Mind: Concerning Computers, Minds, and the Laws of Physics

Penrose's exploration of the mind's role in interpreting quantum mechanics helps ground themes of consciousness and perception in science.

8. Planck, Max.
 - The Universe in the Light of Modern Physics

Planck's work on quantum theory and his reflections on the mind-matter relationship provide early insights into the interplay of consciousness with the material universe.

9. Pribram, Karl H.
- Languages of the Brain: Experimental Paradoxes and Principles in Neuropsychology

Pribram's holonomic brain theory suggests that the mind operates similarly to quantum processes, aligning with the theme of consciousness as a field of potential.

10. Rosenblum, Bruce, and Kuttner, Fred.
- Quantum Enigma: Physics Encounters Consciousness

This work examines how quantum mechanics leads directly to questions about consciousness and reality, bridging scientific discovery and philosophical inquiry.

11. Tart, Charles T.
- The End of Materialism: How Evidence of the Paranormal is Bringing Science and Spirit Together

Tart discusses the interface of science and mysticism through phenomena that defy conventional materialism, adding depth to themes on spirituality's place in modern science.

12. Tolle, Eckhart.
- The Power of Now: A Guide to Spiritual Enlightenment

Although not a scientific text, Tolle's reflections on presence and consciousness contribute to the exploration of the observer's role in creating reality, resonating with quantum theories.

13. Wheeler, John Archibald.
- Geons, Black Holes, and Quantum Foam: A Life in Physics

Wheeler's idea of "participatory universe" and "it from

bit" highlight the interactive nature of reality, framing existence as participatory and consciousness as central to the quantum field.

14. Wilber, Ken.
   - A Brief History of Everything

Wilber's work synthesizes spiritual, psychological, and philosophical ideas, useful for understanding the unity between science and spirituality.

15. Zohar, Danah, and Marshall, Ian.
   - The Quantum Self: Human Nature and Consciousness Defined by the New Physics

Zohar and Marshall explore how quantum theory can redefine consciousness and identity, supporting the theme of the self as part of a unified field.

ACKNOWLEDGEMENTS

The journey of creating Quantum Mysticism: The Spiritual Implications of Modern Physics has been one of deep reflection, inspiration, and the joy of connecting with the mysteries that bridge science and spirituality. This book would not have been possible without the guidance and support of numerous individuals whose insights, encouragement, and wisdom enriched every step of the process.

I am deeply grateful to the scientists and thinkers who have dedicated their lives to expanding our understanding of the universe and consciousness—especially the works of David Bohm, Fritjof Capra, and Amit Goswami, whose perspectives helped shape the ideas presented in this book. Their willingness to challenge the boundaries of knowledge and open up dialogues between science and spirituality inspired much of what is written here.

A special thank you goes to my publisher, Irene Minds, for believing in this project and supporting its vision. I am also grateful to friends, family, and colleagues who have provided valuable feedback, encouragement, and support along the way.

Finally, my heartfelt thanks to all seekers and readers who open themselves to the beauty of the unknown and the unity of all things. It is to you that this work is dedicated. May this book serve as a guide, a companion, and a source of inspiration as we explore together the infinite mysteries of existence.

COPYRIGHT INFORMATION

Quantum Mysticism: The Spiritual Implications of Modern Physics
© 2024 by Dr Bhaskar Bora
Published by Irene Minds

All rights reserved. No part of this book may be reproduced, distributed, or transmitted in any form or by any means, including photocopying, recording, or other electronic or mechanical methods, without the prior written permission of the publisher, except in the case of brief quotations embodied in critical reviews and certain other non-commercial uses permitted by copyright law. For permission requests, write to the publisher at:

Irene Minds
Email: bora.dr@gmail.com

DISCLAIMER

This book, Quantum Mysticism: The Spiritual Implications of Modern Physics, is intended for informational and contemplative purposes only. The ideas and concepts presented are interpretations of scientific and philosophical insights and are not intended to serve as definitive scientific explanations or spiritual doctrines. The author is not providing medical, scientific, psychological, or spiritual counselling, and readers are encouraged to explore these topics independently or consult professionals in respective fields for guidance.

While efforts have been made to provide accurate and researched content, the author and publisher assume no responsibility for any errors, omissions, or interpretations of the material. The perspectives offered here are meant to inspire reflection and exploration, and any decisions or actions taken by readers based on the content are solely their responsibility.

Thank you for embarking on this journey of discovery and reflection.

www.ingramcontent.com/pod-product-compliance
Lightning Source LLC
Chambersburg PA
CBHW031616210526
45464CB00004B/1604